Project Management Concepts, Methods, and Techniques

PROGRAM MANAGEMENT TITLES AVAILABLE
FROM AUERBACH PUBLICATIONS, TAYLOR & FRANCIS GROUP

PgMP® Exam: Practice Test and Study Guide, Third Edition
Ginger Levin, and J. LeRoy Ward
978-1-4665-1362-4

Program Management Complexity: A Competency Model
Ginger Levin, and J. LeRoy Ward
978-1-4398-5111-1

Implementing Program Management: Templates and Form Aligned with the
Standard for Program Management—Second Edition (2008)
Ginger Levin and Allen R. Green
978-1-4398-1605-9

ESI International Project Management Series
Series Editor
J. LeRoy Ward, Executive Vice President
ESI International, Arlington, Virginia

Project Management Concepts, Methods, and Techniques
Claude H. Maley
978-1-4665-0288-8

Project Management for Healthcare
David Shirley
978-1-4398-1953-1

Managing Web Projects
Edward B. Farkas
978-1-4398-0495-7

Project Management Recipes for Success
Guy L. De Furia
978-1-4200-7824-4

A Standard for Enterprise Project Management
Michael S. Zambruski
978-1-4200-7245-7

Determining Project Requirements
Hans Jonasson
978-1-4200-4502-4

The Complete Project Management Office Handbook, Second Edition
Gerard M. Hill
978-1-4200-4680-9

Other ESI International Titles Available
from Auerbach Publications, Taylor & Francis Group

PMP® Challenge! Fourth Edition
J. LeRoy Ward and Ginger Levin
978-1-8903-6740-4

PMP® Exam: Practice Test and Study Guide, Seventh Edition
J. LeRoy Ward
978-1-8903-6741-1

Project Management Concepts, Methods, and Techniques

Claude H. Maley

CRC Press
Taylor & Francis Group
Boca Raton   London   New York

CRC Press is an imprint of the
Taylor & Francis Group, an **informa** business
AN AUERBACH BOOK

CRC Press
Taylor & Francis Group
6000 Broken Sound Parkway NW, Suite 300
Boca Raton, FL 33487-2742

© 2012 by Taylor & Francis Group, LLC
CRC Press is an imprint of Taylor & Francis Group, an Informa business

No claim to original U.S. Government works

Version Date: 20120420

International Standard Book Number: 978-1-4665-0288-8 (Hardback)

Visit the Taylor & Francis Web site at
http://www.taylorandfrancis.com

and the CRC Press Web site at
http://www.crcpress.com

Contents

Preface

OVERVIEW

Project Management Concepts, Methods, and Techniques focuses on the *why, what,* and *how* of the management of projects, and seeks to develop and improve the skills required for managing ongoing and future projects:

- Comprehend project management concepts and their significance
  - Understand how to meet business goals through effective project management
  - Learn how to deliver and realize projects according to stakeholders' expectations
- Develop core project management skills to contribute to the achievement of strategic objectives of your organization
  - Apply the techniques and tools to plan and schedule projects
- Learn the major supporting skills necessary to apply project management in a multidisciplinary and cross-functional environment
  - Learn how to manage projects with external providers
- Discover the performance obstacles to effective project management and how to overcome them to produce positive results
- Ascertain the communication and relationship management techniques that are keys to project success

Each topic is structured in a balanced blend of detailed explanatory texts, rich diagrams and graphics for a complete understanding of the subject. Illustrated examples and exercises, with worked-through answers, strengthen the learning objectives and comprehension.

This book, completely aligned with the Project Management Institute Body of Knowledge, is an ideal platform for learning and/or consolidating project management knowledge, for immediate implementation in the work environment.

TARGET AUDIENCE

The book addresses all readers who participate directly or indirectly in the management of projects. The audience includes project managers and project leaders, line managers, executives, and internal and external project providers, suppliers of projects, and those who acquire external project services.

The book is highly valuable for all professionals who wish to review knowledge of the major concepts, methods, and techniques of project management and gain practical knowledge and effective application of project management, in particular issues related to the organization, planning, realization, and control of projects.

Selection of Programs/Projects

Industry	Key Project	Corporation
Automotive	Process methodology to reduce design & manufacturing cycle	Renault Automobile
Banking	Back office information system for consolidation	Nova Ljublanska Banka
Distribution	Turnkey warehousing and logistics	Cartier-Interflamme
Energy	Engineering bid management organization and process	Areva
Engineering	Land and estate management development	Seranillos
Food & Beverages	Turnkey processed food production plant	Ganados de Salamanca
Insurance	Automotive policy handling system	GAN
Manufacturing— discrete and heavy	Engineering and manufacturing of heavy machinery	Caterpillar
Marketing Services	World-wide market analysis systems	IMS
Oil and Gas	Management of Projects Methodology for Exploration & Production	BP
Pharmaceutical	Global new product launch	Organon
Public Health	Project management office for national health insurance electronic payment	CNAM
Shipping and Transportation	Container logistics	OCL

Technology and information systems	Organizational changes to evolve to added-value solutions provider	Hewlett-Packard
Telecommunications and Telephony	Organizational changes to evolve to added-value solutions provider	Ericsson
Siemens - Iskratel		

Consulting, Education and Training Themes	**Topics**
Management of change by programs	Change management; solution management; managing programs; management by projects; strategic project management
Management of projects	Concepts, methods, and techniques; project risks; planning & delivery; PMP Preparation
Leadership, motivation, and team management	Effective project leadership; effective communication and relationship management; managing teams

Claude H. Maley

Claude Maley is Managing Director of Mit Consultants, a consultancy and education practice servicing international clients in management of change, and chairman of a business solutions company. He started his career as a systems engineer with IBM, after reading estate management and building construction at the London School of Building, and has held various management positions for international organizations and companies.

His functional management and consulting experience with major corporations such as ABB, Alcatel, Areva, BP, Cadbury-Schweppes, Cartier, Caterpillar, Cisco, Ericsson, GE, Hewlett-Packard, IMS International, Motorola, Organon, Overseas Containers Limited, Pechiney, Renault Automobile, and Siemens, to name but a few, has spanned more than forty years in engineering, production and manufacturing, distribution, transportation, and marketing services sectors. This has exposed him to a variety of situations, which have all forged a deep understanding of the issues governing the management of change by projects.

In a professional career spanning forty years, Claude has held responsibility for a significant number of projects. These have involved internal organizational projects, and external commercial projects, ranging in duration from one month to five years, and in budget from $50,000 to $500 million. Claude has also participated in many projects as team leader of subsystems, and he has served on numerous occasions as an external advisor to lead project managers.

Claude is a PMP and professional speaker, instructor, and lecturer in topics ranging from general organizational, program and project management to sales and marketing, leadership, and motivation. In the practice of his consulting and education profession he has worked with more than eighty different nationalities in more than sixty nations in all continents. Claude is fluent in English, French, Spanish, and Italian.

Claude is author of educational courses and papers on business solutions, management of change by projects, organizational management and leadership, and is a member of the International Project Management Association and the Project Management Institute.

claude.maley@mitconsultants.com

1

Introduction to Project Management

1.1 CHAPTER OVERVIEW

This chapter introduces the fundamentals of project management and the growing need for better project management. It also describes the key elements of the project management framework, including project stakeholders.

We will review the basic terms and concepts that underlie the more detailed methods described in subsequent chapters, explain the constituents of projects, and examine the constraints of projects.

We will consider the role of the project manager, as well as the activities that project managers undertake. We will also assess the skillset project managers need to fulfill their responsibilities. Finally, the chapter will describe the role of professional organizations like the Project Management Institute (PMI) and the project management knowledge areas.

Reference is made throughout to the Project Management Institute, *A Guide to the Project Management Body of Knowledge: PMBOK® Guide, 4th Edition.*

1.2 PROJECTS AND THE BUSINESS ENVIRONMENT

Today's corporations, small-medium companies, governments, and non-profit organizations have recognized that to be successful, they need to be conversant with and use modern project management techniques. In the business environment, individuals are realizing that to remain professionally competitive, they must develop skills to become effective project managers and valuable project team members. Understanding the concepts of project management will help each employee to participate in projects and improve project performance on a daily basis.

Projects come in many forms. There are traditional major projects from heavy engineering industries, such as aerospace, construction, civil works, shipbuilding, transportation, and exploitation of natural resources and energy. These are significant, involving large, dedicated teams and requiring large funds; they are realized over a long period of time, abiding to strict legal, health, safety, and environmental requirements, and calling for the collaboration of several organizations, from owners to suppliers. In the common business environment, most projects are endeavors of engineering and construction to build new facilities, maintain existing facilities, develop and produce launches and/or new manufacturing processes, introduce supply chain management systems, and/or implement new technologies or computer systems, as well as sales remuneration systems or training programs. Outside of the business arena and into the social environment, projects include organizing events in the community or even going on holiday.

Corporations that consistently succeed in managing projects align them with business goals and clearly define what needs to be achieved. They ensure that project management concepts, methods, and techniques are available and utilized.

Companies that perform highly in project delivery focus on important measurements and apply them to all projects. These include achieving business benefits, returns on investment, and creating value.

Furthermore, successful companies recognize that experienced and highly skilled project managers are essential to project success. They also know that an effective project manager needs to be business oriented and understand the strategic, financial, organizational, and commercial issues and challenges. Above all, they seek those project managers who possess strong communication and interpersonal skills. Companies that outclass in project management will retain their project managers by providing them with career opportunities, training, and mentoring.

1.2.1 Key Project Management Concepts

Project management is the discipline of planning, organizing, and managing resources to bring about the successful completion of specific project goals and objectives.

The primary challenge of project management is to achieve all of the project goals and objectives while honoring the established and agreed-upon project constraints. Typical constraints are scope, time, and budget.

These must be extended to include the business benefits, the risks, and the quality of the project delivery.

Project Management is "the application of knowledge, skills, tools, and techniques to project activities to meet project requirements" (PMI).

Note that the word "requirements" is an often-used word and has many interpretations. For the purpose of this book, "requirements" will refer to the project constraints (discussed further in Section 1.6, The Triple Constraint).

Project managers must not only strive to meet specific scope, time, cost, and quality goals of projects, they must also facilitate the entire process to meet the business needs and manage risks.

1.2.2 Terminology

Projects are different from operational work. A project is a finite endeavor undertaken to meet particular goals and objectives, to bring about beneficial change or added value. Projects change things and are an integral part of a development effort to achieve business benefits.

Operational work and processes, on the other hand, are continuous and repetitive, and exploit the assets of the company to meet operational goals. An operation thus exploits the result of a project to realize these benefits.

The finite characteristic of projects stands in contrast to operations or processes. In practice, the management of these two systems is quite different, and as such requires the development of distinct technical skills and the adoption of separate management.

A Project

As stated by PMI, a project is "A temporary endeavor undertaken to create a unique product or service." This can also be stated as any undertaking with a defined starting point and defined objectives, by which completion is identified. In practice, most projects depend on finite or limited resources and funding, by which the objectives are to be accomplished.

A Program

"A group of related projects managed in a coordinated way to obtain benefits and control not available from managing them individually" (PMI). Programs are managed by program managers or program directors who manage the work of the individual project managers, and the sponsors might come from several business areas.

1.2.3 Characteristics of Projects

A project is temporary. A project has a definite beginning and a definite end. The time window of the project is part of the overall product life cycle.

A project has a unique purpose. Every project should have well-defined objectives that are aligned to the business benefit. The project result is destined to be exploited.

A project requires resources. Work is performed to achieve the project goals and requires resources. These include human resources, materials, and equipment. The project will also utilize services and corporate assets. Resources are provided by the performing organization or from external sources to achieve the unique purpose. The resources used for the work performed will drive the funding required for the project. Money, as such, is not a project resource, but a facilitator to fund the resources of the project.

A project should have a sponsor. Projects need governance. Projects have many interested parties or stakeholders, but an individual or a group must take the primary role of sponsorship. The project sponsor usually provides the direction and funding for the project.

A project involves uncertainty. Projects create change in an environment of change. It is thus evident and obvious that projects will be performed in the future in an environment of unknowns and uncertainties. The uniqueness of a project highlights the difficulty at the initiating phase in defining the project's objectives clearly, determining the required funding and estimating the duration to complete.

A project uses progressive elaboration. Projects can only be defined in general terms when they are launched. Only through the analysis and realization steps will the specific details of the project become apparent. Therefore, projects need an incremental development approach. Initial project documents are updated progressively with more detail based on new information acquired throughout the project phases.

1.2.4 Projects in the Business Environment

<u>Projects as Agents of Change</u>

Organizations need to maintain, sustain, and grow their business. Organizations that are self-sufficient, efficiently organized, and suffer no competition or external pressure are rare or even may not exist (Figure 1.1).

Organizations have business goals and cannot remain stagnant. They need to constantly measure their internal performance, understand their market and external environments, and take appropriate action to meet their goals. As illustrated below, the organization needs to constantly address its internal and external influences to sustain its business operation and promote innovation and continuous improvement. Executives and operational managers must choose between proactive and reactive decision-making depending on the source of the influences.

Organizational effectiveness has to be continually reviewed to ascertain changes that have to be made, and whether these changes address ongoing operational performance or seek to adopt a new way of operating as a company. The funding of the resulting decisions will be allocated either to an operational expenditure, usually the yearly financial budget, or to a capital expenditure, which would not be restricted to a calendar boundary.

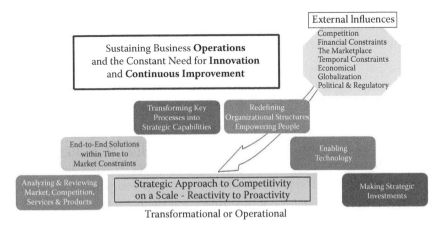

FIGURE 1.1
Sustaining business operations

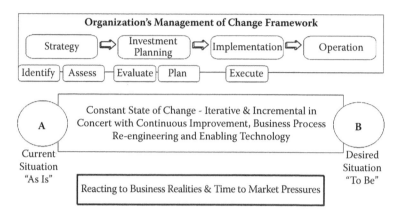

FIGURE 1.2

Management of change framework

Organizations will create change in their operations, be it through a strategic transformational initiative to create new performance levels and/ or to ensure that operational performance levels are efficient and profitable. The change will transform the organization from its current situation to a desired and new situation.

The Figure 1.2 illustrates the overarching time frame for change, and calls upon continuous improvement and business process engineering to achieve the desired results. Note that only in the new and desired operational situation can the business benefits be measured.

Projects and programs respond to the demand for change and will be driven by either transformational or operational business needs. Projects and programs will contribute as the principal instruments for change and directly create the business value (Figure 1.3).

1.2.5 Strategic and Tactical Projects

Projects intended for launch will have, as their source, either a strategic initiative, usually with capital expenditure funding, or an operational need, utilizing the operational expenditure funding.

It is important to assess the extent of the change compared to its funding to determine the nature of the project and the intended business benefits. Figure 1.4 highlights the demarcation between evolutionary changes, usually made on an existing operation, and revolutionary changes that create innovation.

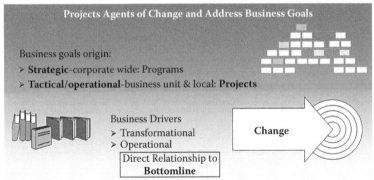

FIGURE 1.3
Projects as agents of change

Revolutionary changes are the strategic projects and must be aligned to the corporate strategic intents. Evolutionary changes are the tactical projects and are for the most part driven by the operational entity or entities that require sustaining and maintaining ongoing operations.

1.2.6 Projects and Programs

As noted previously, a project is "A temporary endeavor undertaken to create a unique product or service," whereas a program, according to PMI,

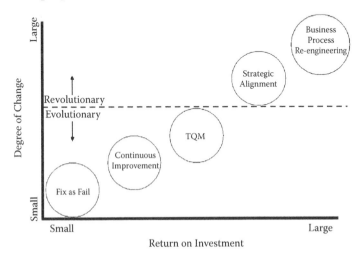

FIGURE 1.4
Evolutionary and revolutionary change

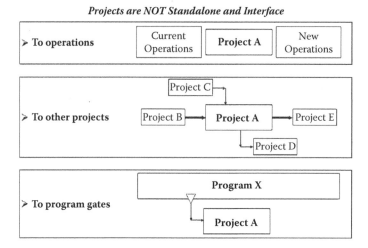

FIGURE 1.5
Projects are not standalone

is "A group of related projects managed in a coordinated way to obtain benefits and control not available from managing them individually." Projects produce deliverables whereas programs meet business needs. Single and stand-alone projects will of course produce deliverables that meet the business goals.

Programs must be designed to perform projects early in the cycle to allow quick business value while ensuring that infrastructure projects facilitate meeting these goals.

Projects have the characteristic of not only forming part of a program, but they can be stand-alone or interconnected to other ongoing projects (Figure 1.5).

1.3 THE LIFE CYCLE

1.3.1 What Is a Life Cycle?

A life cycle refers to the phases that a product or service undergoes: research and development, release, growth, expansion, maturity, saturation, decline, and retirement. The life cycle covers the lifespan of a corporate product or service destined for the marketplace, or of an organizational asset or process destined for use internally by the corporation.

1.3.2 The Product Life Cycle

The product life cycle usually refers to a marketed product or service, and it spans the period that begins with the initial product specification and ends with the withdrawal from the market of both the product and its support.

When the product life cycle refers to an organizational asset or process, it covers a period from development, through testing and handover to operational usage. Such assets or processes, through their usage, provide operational capabilities for the corporation. Over time these assets or processes will evolve and generate modifications and/or improvements until they become either obsolete or too costly to operate, leading to their retirement and/or replacement.

The product life cycle is characterized by a period of ownership followed by a period of operations (Figure 1.6).

The ownership period concerns the acquisition of the product, service, asset or process. The period ends when release is made to operations to exploit. Ownership, as the word infers, is cost-driven. Funding will either be an operational expenditure (OPEX) or a capital expenditure (CAPEX).

The operations period exploits the product, service, asset, or process. The period ends with retirement or obsolescence. The operations period is benefit-driven. These benefits—usage of assets and processes or generated revenues from products and services—are offset against the cumulative ownership costs and ongoing operational costs.

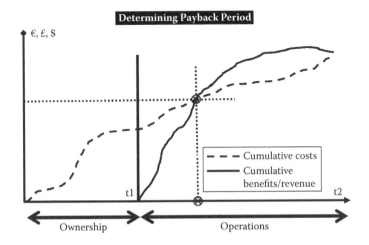

FIGURE 1.6
Determining payback period

1.3.3 The Project Development Life Cycle

The project development life cycle explicitly refers to the ownership period.

The project development life cycle focuses on the research and development, realization, and release of products, services, assets, and processes; it addresses the needs, opportunities, and issues of the performing organization.

The project development life cycle consists of several phases to conceptualize both the problem and the solution, and to materialize the solution for operational use. The phases can subsequently be broken down into stages.

The cycle can be compared to a problem-resolution exercise, as illustrated in Figure 1.7

The linear and chronological approach in Figure 1.7 forms the backbone of a project development life cycle. However, as shown in Figure 1.8, Figure 1.9, and Figure 1.10, different iterative and cascading models exist to adapt to different dynamic project environments.

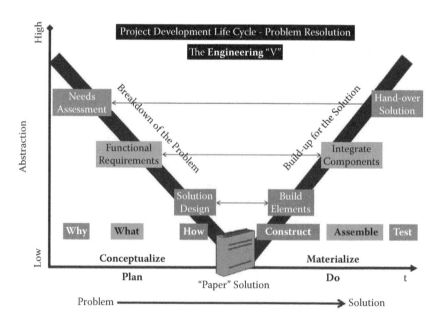

FIGURE 1.7

Project life cycle—engineering V

Project Life Cycle: Waterfall Model

Requirements

Design

Realization &
Unit Testing

Integration &
System Testing

Operation &
Maintenance

FIGURE 1.8
Project life cycle—waterfall model

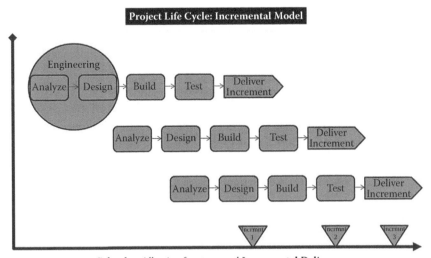

Project Life Cycle: Incremental Model

Engineering

Analyze → Design | Build → Test → Deliver Increment

Analyze → Design → Build → Test → Deliver Increment

Analyze → Design → Build → Test → Deliver Increment

Incrmnt 1   Incrmnt 2   Incrmnt 3

Calender: Allowing for staggered **Incremental Delivery**

FIGURE 1.9
Project life cycle—incremental model

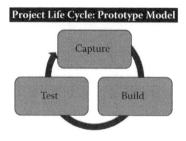

FIGURE 1.10
Project life cycle—prototype model

1.3.3.1 Characteristics of Project Phases

The key essence of the project phases is to provide a beneficial solution to the organization, as uncertainties and unknowns are addressed. Scope management will be at the core of this and will be discussed at length in Chapter 2 of this book.

Each project phase produces one or more results—deliverables. A deliverable is a work product, such as a needs assessment, a solution design, or a functional proof of concept.

The phase sequence defined by most project development life cycles involves some form of transfer or handoff, for example, requirements to design, construction to operations, or design to manufacturing.

The conclusion of a project phase is generally marked by a management review of both project performance and results to determine if the project should continue into its next phase. These reviews are called phase exits or stage gates.

The project development life cycle defines the nature of work to be performed in each phase and the profiles of resources involved in each phase. Cost and staffing levels are the lowest at the start, rise rapidly midstream and are at the highest in the later phases, then drop rapidly as the project nears completion. At the start of the project, risk and uncertainty are at their highest and the probability of successfully completing the project is at its lowest.

Finally, the project development life cycle general structure should progress through the same five phases, as shown in Figure 1.11.

| Origination | Initiation | Planning | Execution | Close Out |

FIGURE 1.11
Project life cycle—PMI phases

PRACTICE EXERCISE

> Reflect on a project that you have been involved in and consider whether it went through the five phases. What happened in each phase? Were any phases left out? If yes, were there things that your organization could have done better if they had followed the five-phase approach?

1.4 PROJECT GOVERNANCE

1.4.1 Overview

Project governance refers to the design and implementation of formal principles, structures, processes, and communications that need to exist for a successful project program or portfolio of projects. Project governance is the framework within which decisions are made on projects (see Figure 1.12).

Project governance:

- Establishes the relationships between all internal and external groups involved in the project
- Describes the flow of project information to all stakeholders
- Ensures the appropriate review of issues encountered within each project
- Ensures that required approvals and direction for the project is obtained at each appropriate stage of the project

FIGURE 1.12
Project governance framework

Project governance is initiated under the umbrella of corporate governance, appoints a governing body for a project, and makes a significant contribution to addressing the following concerns:

- The business case—by stating the business benefits to achieve and the objectives of the project, specifying the in-scope and out-of-scope elements
- Identified stakeholders—with an interest in the business case and the project
- The business-level requirements—as agreed by all stakeholders
- A stakeholder communication plan
- Agreement on the project deliverables—and a mechanism to assess the compliance of the completed project to its original objectives
- An appointed project manager
- Clear project roles and responsibilities assignment—providing a basis for decision-making
- A published project development plan—updated throughout the project phases
- A clear set of processes—to manage risks, issues, scope changes, and quality
- A project review and progress-reporting process—including quality and documentation assessments
- A central repository for the project including a glossary of project terms

1.4.2 The Project Board

The project board is the forum where senior management representatives of the organization come together to make decisions and commitments to the project. The project sponsor chairs the board.

The project manager is *not* a member of the project board. The board will request board meeting attendance to provide progress updates and assist in making decisions in a timely manner.

The board gives the project manager authority to manage the project within the tolerances set by the project board.

1.4.3 The Sponsor

The sponsor is the "governor" for a project and therefore provides the link between corporate and project governance. Ideally, the sponsor should be a member of the corporate executive.

"Project sponsors are the route through which project managers directly report and from which project managers obtain their formal authority, remit and decisions. Sponsors own the project business case. Competent project sponsorship is of great benefit to even the best project managers" (Association for Project Management).

The sponsor is ultimately responsible for the project and has to ensure that the project delivers corporate value and addresses the needs of the business. The key specific responsibilities of the sponsor are:

- Organize and chair project board meetings
- Approve the business case
- Authorize project expenditure
- Set phase tolerances for phase reviews
- Brief corporate and program management about project progress
- Review project progress and its alignment to the business benefits and other constraints
- Recommend and/or approve corrective actions when tolerances are likely to be exceeded
- Approve the end project report
- Approve the sending of notification of project closure to corporate and program management

1.4.4 Strategies for Effective Governance in Projects

Establish a governance structure with clear accountabilities and responsibilities for key-decision making.

Roles and responsibilities of project managers and project leadership should be clearly described and documented.

Introduce or standardize the processes, which is important in establishing good governance. Organizations such as PMI provide the standards for project management, which can be applied to any project environment to introduce good governance.

Organizations can also introduce processes to align their corporate strategy with the project strategy and promote good governance. These processes are to be monitored and evaluated to determine whether the objectives are met.

Better governance in projects leads to stakeholder satisfaction as it will help ensure that projects deliver the expected business value.

1.5 BUSINESS DRIVERS AND BUSINESS NEEDS

1.5.1 Establishing the Business Needs and Initial Scope

Projects must originate with a clearly stated business intention. They must also demonstrate how they will create sustainable value to the organization.

A formal origination process is important because it creates a structured approach to assess all proposed initiatives, avoids allocating resources on lower priority projects, and focuses the organization on projects that will deliver the greatest value.

The purpose of origination is to provide a mechanism for recognizing and identifying potential projects within the organization, to evaluate projects proposed for the next investment cycle, and to reach a consensus on projects to be selected. During this phase, the strength of a project's business case is tested, and the viability of the proposed solution is explored against the company's strategic plan and budget guidelines.

Other factors to take into consideration include legislative restrictions, regulations, and health, safety and environment (HSE) requirements.

The key components of the origination process are:

1. The project must provide sufficient information about the viability of the project's business case and the feasibility of its proposed solution.
2. Projects must be assessed, evaluated, ranked, and prioritized using a consistently applied methodology.
3. The process must consider the project's fit with the organizational mission and strategic plan.

1.5.1.1 The Business Case

The development of the business case is a mandatory step prior to any project launch. The more the business need is pertinent to sustainability and growth, the more comprehensive the business case needs to be. Business cases can range from a "one-pager" to a full and all-inclusive documented analysis of current and future situations that meet business needs. All business cases, however, must include the following:

1. The business problem—providing a thorough and objective presentation of a business issue that requires a timely solution.

2. Details of the project's fit with organization's mission—describing how the expected outcome of the project supports and aligns to the organization's strategic or operational intent.

3. Anticipated benefits—providing qualitative and quantitative estimates of the expected benefits to the organization. These benefits are to be qualified in terms of their impact on the organization's operational performance.

4. Cost/benefit analysis—providing a comprehensive and sincere assessment of both expected benefits and anticipated costs throughout the total product life cycle.

5. Funding and its sources—identifying the ownership costs required for the project development and the operational costs for the utilization of the end-product or service.

1.5.1.2 Achieving Business Alignment

In the origination stage, many projects will contend for limited funding and resources. Each project is to be assessed for its alignment to the strategic or operational corporate goals. The major steps below provide a road map to ascertain the project's alignment to the business.

- **Current Situation Analysis "As Is"**—assess the ongoing and current operation by analyzing the business processes with their owners. Establish the boundary of the project and identify the issues to address and resolve.
- **Desired Future State Definition "To Be"**—describe the desired future state, perform a gap analysis, identify a roadmap to get to the desired future state, and build a portfolio of key projects backed by key financial data including return-on-investment (ROI) analysis and net present value (NPV). Define a set of objectives.
- **Interaction with the Business Strategy**—define and implement a formal and purposeful communication program with business process owners and line-of-business decision makers.
- **Develop an Organizational Infrastructure Model**—establish a model describing the relationship between organizational infrastructure, business processes, and information management needs.

1.5.1.3 The Project Proposal and Proposed Solution

The project proposal and solution respond to the business case requirements. At the early stages of origination, the proposal and solution are high-level guidelines and contain a large number of assumptions.

The proposed solution should offer an approach to solving the business needs, and present the alternative solutions considered and evaluated. The proposed solution should address only the issues identified by the business problem and project objectives should be explicitly stated (see below). The solution should demonstrate how the end-product or service will be integrated into the existing business in financial terms (cost/benefit), offer organizational alignment (processes and training), and provide a high-level schedule for project development and handover to operations, clearly highlighting milestones and assumed dates.

The solution must include the funding and resources requirements for the project, and the estimated costs for operational support/maintenance and other recurring costs.

An organizational impact analysis must accompany the solution proposal and should describe the impact of the project on people, processes, and technology.

1.5.2 The MOST Model

The MOST model provides a rational framework for investments and activities, and it establishes accountability and efficiency across the organization. The model is used at all levels of a corporation and is specifically effective for projects. MOST sets the project's framework and ensures there is coherence between the different steps of the model (Figure 1.13).

The four components of the model are:

Mission: states the purpose of the project and its "reason for being," and declares the goals.
Objectives: are the measurable quantification of the goals.
Strategy: is the approach to achieving the objectives.
Tactics: are the detailed methods for achieving each strategy.

The **mission** statement provides a backdrop for why the project exists. It is both a rallying point for all project players, as well as a clear statement of intent for all those affected by the project as a whole. A mission statement guides all planning across the different components of the project, and

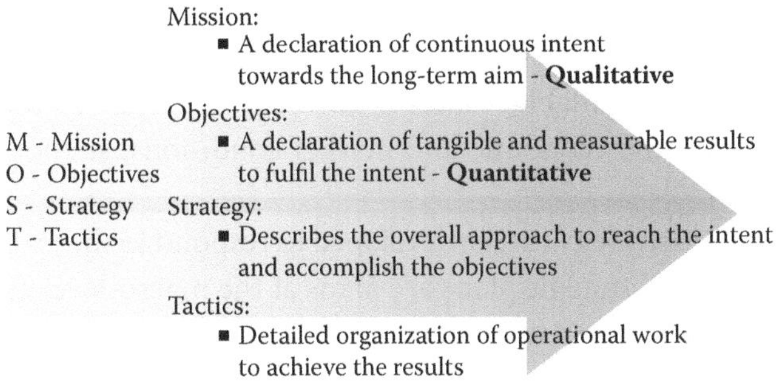

Mission:
- A declaration of continuous intent towards the long-term aim - **Qualitative**

Objectives:
M - Mission
O - Objectives
S - Strategy
T - Tactics
- A declaration of tangible and measurable results to fulfil the intent - **Quantitative**

Strategy:
- Describes the overall approach to reach the intent and accomplish the objectives

Tactics:
- Detailed organization of operational work to achieve the results

FIGURE 1.13
The MOST model

indicates the purpose and activities of the project, for example, "to design, develop, manufacture, and exploit *specific product/service* for *use/sale* to meet the *identified needs* of specified groups via *certain means* in *particular areas.*" Mission statements will refer to goals to be met. These are major results or targets to achieve, are sometimes nebulous, and are not specific enough to be measured.

Objectives state what needs to be done to achieve the mission and its goals, in the medium/long term and according to what metrics. An objective clearly defines the parameters and measurements for success.

Objectives should relate to the expectations and requirements of all the major stakeholders, including employees, and should reflect the underlying business reasons for launching the project. These objectives could cover growth, profitability, offerings, and markets.

Objectives are more specific than goals. Whereas goals are broad, objectives are narrow. Whereas goals are general intentions, objectives are precise. Whereas goals are intangible, objectives are tangible. Whereas goals are abstract, objectives are concrete.

Objectives define the quantitative criteria for a mission and its goals. As such, they need to be complete, realistic, and readily measurable. They are also the foundation for a project's action plans, or strategies.

Objectives are

- Specific
- Measurable
- Achievable
- Realistic
- Time-based

A **strategy** describes the approach the project will pursue to achieve the objectives. Strategies should make clearly explicit the project's needs and assumptions about available resources and critical success factors, including people, funding, economic and market conditions, supporting technologies, etc.

More than one strategy can be developed to respond to the project objectives. The project's strategic plans are plans at the macro-level and as such set the high-level actions to be performed. This is the case for programs, which consist of a series of projects.

Tactics are the actual activities, tasks, and specific deliverables the project has determined will support the achievement of its strategies. If more than one strategy exists, different tactics will be established for each. Tactics detail the planning and budgeting of the activities to support the strategies.

Tactics are the end result of all the planning steps. If the process has been thoroughly implemented it should be possible to trace each tactic to its strategy and the originating objective, to the project's major goals, and finally to the project mission.

1.5.3 Project Stakeholders

Stakeholders are the people involved in or affected by project activities and include the project sponsor, project team, support staff, customers, users, suppliers, and even opponents of the project. These stakeholders often have very different needs and expectations.

This topic will be fully explained and developed in Chapter 2.

1.6 THE TRIPLE CONSTRAINT

1.6.1 Definition

The triple constraint is a concept that is applicable to any kind of project. It is used to illustrate how key vectors of the project are interconnected, and how changing one has an effect on the others. As shown in Figure 1.14, it brings together the three major constraints: scope, cost, and time, represented as a triangle. As any one side changes length, the other two sides have to change in synchrony.

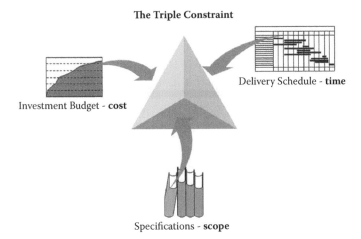

FIGURE 1.14
The triple constraint

Project scope is defined as the work that must be performed to deliver a product, service, or result with the specified features and functions (*PMBOK Guide*).

The scope of any project is a key component to its success. The scope defines the project expectations and its contents and sets the project's boundary.

As the scope contents evolve, either or both cost and time need to be adapted. Cost will be modified because a different resource mix is needed for the activities. Time will be modified because of the variance of the work volume to be performed. If time cannot change, cost will be adjusted by an increase of resources in order to meet the time constraint.

Project "inflation" (scope creep) happens when the requesting organization wants to increase the scope of a project without changing the cost or time. Scope changes can only be incorporated with resulting changes to cost and/or time.

Controlling the three constraints leads to resources management and quality control. The project scope management plan is the key central document. This document describes how the project scope will be defined, developed, and verified, and how the work breakdown structure will be created and defined. It also provides guidance on how the project scope will be managed and controlled by the project management team. The project scope management plan can be informal and broadly framed, or formal and highly detailed, based on the needs of the project (*PMBOK Guide*).

The triple constraint has, for many years, been promoted by PMI; however, recently it has been expanded to address wider constraints.

1.6.2 Extended Issues to the Triple Constraint

The triple constraint has its limitations as it does not respond to and address the wider project constraints. There are at least three additional vectors that have to be considered in the concept of project constraints:

1. The business benefit which drives the purpose of the project. All projects exist to respond to a business requirement.
2. Projects by the nature of things are performed in the future in an environment of unknowns and uncertainty. Thus, risk must be addressed and managed.
3. The project results have not only to respond to the business need but must imperatively correspond to the quality of its contents. This goes beyond just fulfilling the scope.

To these extended constraints, other constraints must be integrated, such as organizational and legal requirements and HSE. These are often included in the scope constraint. Resource constraints are part of both cost and time constraints.

The project constraints (Figure 1.15) can thus be represented as a cube, in which each constraint "face" plays an influence on the other five "faces" (Figure 1.16).

The six faces of the cube compete with each other, and it is rare that all constraints fit. The project manager must establish the priority with the sponsoring group and the extent of the trade-offs. Business benefits and the resulting scope need to be assessed as the project progresses through the development phases. Milestones are to be set at phase exits to determine the changes to be incorporated at each end of phase review. This needs to be clearly established in the project scope management plan.

1.6.2.1 Business Benefit

Business benefit is intimately associated with the purpose of the project and is core to the overall ownership and operational benefit/cost of the product life cycle.

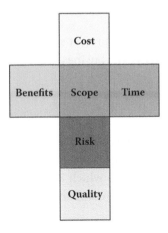

FIGURE 1.15
The six constraints

The project manager must be conversant with the business reasons driving the project and must establish with the sponsoring group the priority to be given to this constraint.

1.6.2.2 Scope

When scope is increased or changed constantly, then cost and time must be adapted. Failure to adjust may increase the risks to the project, decrease the quality of the final product or service, and therefore fall short of the business benefit.

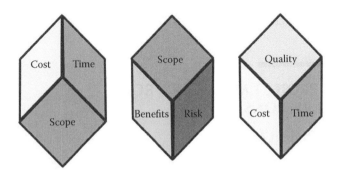

FIGURE 1.16
The "cube"—multifaceted constraints

1.6.2.3 Risk

Risk management is initiated at project selection. The analysis, planning, and control of risks influence the business benefits, the project scope, cost, time, and/or quality.

1.6.2.4 Cost

The project cost is subordinated to the overall ownership and operational costs of the product life cycle. The project cost is often imposed at the origination of the project. In cases in which the cost budget can be developed and proposed, this will always be compared to the financial and economic merits of the business benefit.

The project manager must agree with the sponsoring group that if cost is reduced, then the scope of the deliverables has to be reduced with an impact on schedule, or the business benefit, risk, and quality will suffer.

1.6.2.5 Time

Similar to the project cost, the project time constraint is subordinated to the ownership and operational time frame of the product life cycle. Here again, the project time is often imposed at the origination of the project.

The project manager must agree with the sponsoring group that if the project time frame is changed, the schedule is subsequently affected, as well as the cost. Reduction in the project time frame forces a reduction in the scope, giving rise to lower quality and increases risks while decreasing the business benefit.

1.6.2.6 Quality

Quality affects the business benefit. It is compromised when the project scope, time, and cost change or are arbitrarily imposed.

1.7 THE PROJECT MANAGER'S ROLE AND RESPONSIBILITIES

1.7.1 The Project Manager

A boat without a skipper will navigate; however, it may never successfully reach its destination. An organization without management will operate, however, it may never successfully achieve its objectives. Organizational environments differ, from business enterprises to nonprofit organizations and governmental public structures. They will all perform large, medium, and small projects. Project managers are the skippers.

An organization will choose how to establish its management of projects, and give it the visibility and importance it wishes. Those organizations that focus more on day-to-day operational results may often treat the management of projects as a means to sustain and maintain operational performance. In these cases, projects are considered more as continuous maintenance exercises, and there would be few or no recognized project managers. Those organizations that view operational performance as strategic will more often than not recognize the important and official role of project managers and will institute a formal and methodological approach to the management of projects. Those organizations that promote and sell their services, especially in fixed-price contracts, will elevate the management of projects to a high visibility and consequently the position of a project manager is appropriately recognized.

No matter the approach taken by an organization, those individuals who are tasked to manage a project have to perform as project managers for the duration of the project, and require a wide palette of skills applicable to all project environments. Many people are assigned to manage projects; however, it must be remembered that a project manager is a professional in the field of project management.

Projects introduce change in organizations and involve changes within the projects themselves. Project managers must be comfortable leading and handling change. Project managers need to understand the organizations they work in and how products are developed and services are provided. And they must also understand the social, physical, and political environment.

A project manager is the individual accountable for accomplishing the project objectives. The overarching responsibility of project managers is the planning, execution, and closing of the project. That responsibility extends to the wider organization, as project managers have to promote and participate in the organizational readiness of the operation before hand-over.

1.7.1.1 Project Manager Job Description

A project manager can have many different job descriptions, which can vary tremendously based on the organization and the project. Some companies will be specific in these descriptions, whereas others draft a loosely phrased definition. Below is a selection of those descriptions:

- Manages, prioritizes, develops, and implements solutions to meet business needs
- Applies technical, theoretical, and managerial skills to satisfy project requirements
- Builds and maintains an effective and open communication network with sponsors and key stakeholders
- Coordinates and integrates team and individual efforts and builds positive professional relationships with stakeholders and external providers
- Conducts and coordinates business analysis, requirements gathering, project planning, schedule and budget estimating, development, testing, and implementation
- Manages project risks and prepares contingency plans
- Controls implementation activities to fulfill identified objectives
- Institutes and manages the project change control process
- Establishes cross-functional teams for organizational readiness
- Acts as a link between internal users and external providers to develop and implement solutions
- Participates in vendor contract development and budget management
- Responsible for external providers to ensure development is completed in a timely, high-quality, and cost-effective manner
- Provides post-implementation support

PRACTICE EXERCISE

Does your organization have job descriptions for project managers? If yes, review one of these and see if you think there are activities that should be added or removed. If no, discuss this section with your HR department (or the person responsible).

1.7.2 Project Manager's Palette of Skills

As project management is made more visible in an organization and effective project management methodologies are instituted, the project

FIGURE 1.17
The project manager's major functions

manager needs to develop wider hard and soft skills (Figure 1.17). These skills revolve around:

- Organization and management
- Technical
- People and communication
- Administration

1.7.2.1 Organization and Management

Project managers should possess general management knowledge and skills. They should understand important topics related to strategic planning, tactical planning, organizational structures and behavior, operations management, personnel administration, compensation, benefits, career paths, financial management, accounting, procurement, sales, marketing, contracts, manufacturing, distribution, logistics, supply chain management, and health and safety practices.

On most projects, the project manager will need a wide and deep experience in one or several of these general management areas. On other projects, the project manager may delegate some of these areas to a team member, support staff, or even an external provider, while still maintaining the overall responsibility. Even so, the project manager retains accountability for all key project decisions.

1.7.2.2 Technical (Subject Matter Expertise)

The debate as to whether the project manager is to have knowledge, experience, and skills in the technical arena has lasted since projects and project managers have existed.

The "technical" skill is the core "subject matter" skill needed for the project, and it depends on the project. For a wide range of projects covering construction, civil engineering, information management, sales and marketing, biochemistry, financial, legal, etc., rarely would a project manager have the technical subject matter expertise in all the areas. Usually most project managers find themselves in a project environment for which they have had education, exposure, and experience in some of the core key subject matter areas.

For a project manager, the project management skills are the most important. However, both project management skills and technical (subject matter) skills are necessary for success of a project. The technical skills level required mainly depends on the size of the project and the industry. Project management skills are often directly proportional to lack of subject matter expertise availability. The project management skill can balance the lack of the expertise of subject matter to some extent.

It is very important for the project manager to communicate and understand the technicalities of the project with the project team. Small projects often require more technical skills than large projects. In small projects, the project manager is closer to the people performing on the project, and the team expects the project manager to be involved in technical discussions. In large projects, it may be adequate for the project manager to have a broad view of the underlying technical area. As projects become larger, the people and communication skills become more critical.

In some industries, the project manager commands more respect from the team with a similar subject matter background. However, the pitfall is for the project manager to abuse the possessed technical skills by micromanaging or totally ignoring the project management role.

As with all employees, project managers should have the technical subject matter knowledge and skills needed to do their jobs. On smaller projects, technical training may enhance the abilities of project managers to contribute technically, but it is unlikely to improve their management skills, other than being more credible with the project team. On larger projects, there are too many complex technologies for the project manager to master. Technical training that provides breadth may be useful. Acquiring technical knowledge allows the project manager to guide the career development of junior staff and can help senior staff make the transition to management.

Without technical knowledge, the project manager must rely on a technical lead to provide this knowledge. The key questions that project managers should ask include the following:

- What types of technical problems require management?
- Where are they, and who will solve them?
- How will quality and satisfaction be measured?
- What outside resources, if any, can I draw on for assistance?

1.7.2.3 People and Communication

Not enough can be said about the project manager's soft and human relations skills. The project manager is the pivot of an organization initiative, dealing internally with all project members and externally with sponsors, stakeholders, upper and line managers, peers, and external providers.

Achieving high performance on projects requires effective communication, influencing the organization to get things done, leadership, motivation, negotiation, conflict management, and problem solving.

The project manager must hone the leadership skills required for the job so as to understand the greater needs of their stakeholders and organizations, and meet expectations, while engaging and seeking commitment from project members. Project managers need to communicate, lead, negotiate, solve problems, and influence the organization at large. They need to be able to listen actively to all project players, promote new approaches for solving problems, and influence others to work toward achieving the project goals.

Project managers must lead their project teams by providing vision, creating an energetic, participative, and positive environment, and setting an example of appropriate and effective behavior. They must focus on teamwork skills to motivate project members and to develop an *esprit de corps* within the team to achieve effective and high-quality results.

As most projects involve trade-offs between competing goals, it is important for project managers to have strong personality skills to cope with criticism and constant change. Project managers must be open, pragmatic, realistic, flexible, creative, and often patient in guiding the project to its goals.

The most significant characteristics of effective project managers are

- Convey the corporate goals clearly
- Articulate the project directives visibly
- Be inspirational
- Be decisive
- Lead by example
- Be a good communicator
- Be a good motivator
- Be technically competent
- Align team members
- Support team members
- Encourage innovation and creativity
- Challenge top management decisions when necessary

1.7.2.4 Administration

Project managers must focus on getting the job done by paying attention to the details and daily operations of the project.

Project administration functions are different from project management functions. The administration functions cover a host of processes and "desk-work" activities, ranging from project tracking and control to scheduling meetings, issuing minutes, compiling and expediting all the documentation and deliverables for the project, and providing status reports on tasks and deliverables.

The scope of project administration is to collect, collate, monitor, and provide information relating to the management of projects, including work schedules, financial and budgetary data, requirements, and reporting as required.

Project management software tools are of prime importance for data administration ensuring the integrity, consistency, and timeliness of planning, tracking, and reporting. These tools also provide a single focal point for executives, management, and project personnel to access project information, to aggregate individual project plans, and to produce summary status reports.

When project managers have to manage projects and perform administrative functions, they need to establish a robust process for data and information management systems. Projects, which require a great deal of communication and information collection/dissemination, warrant a dedicated project administrator assigned to assist the project manager. Relieving project managers of the administrative burdens allows them to focus on the project. Additionally, project administrators can assist more than one project manager.

Project managers cannot delegate the communication with steering committees, management committees, and sponsorship groups. Nor can they delegate communications with project team members and stakeholders.

The project manager also performs, or oversees, the following administration activities:

- Review project administration procedures, and prepare guidelines, staff instructions, project administration memorandum, and handbooks
- Monitor and evaluate implementation and development performance, using a project performance management system and a project performance report
- Analyze progress reports
- Compile and report project statistics
- Present progress reports, unaudited and audited project accounts, and financial statements
- Monitor project cash flows
- Review and approve key and major documents
- Monitor compliance
- Recruit and staff project resources
- Coordinate training programs
- Participate in the development and implementation of a document tracking and reporting system for the project.
- Participate in the development and implementation of a document control system for the project
- Disseminate and encourage use of evaluation findings at all stages of the project cycle
- Advise on procurement, consulting services, and matters associated with project implementation
- Prepare standard bidding documents
- Procure and negotiate with external providers
- Produce, issue, and amend standard contracts for external services

- Monitor contract awards, disbursement, and other project administration indicators
- Participate in the development and implementation of a project correspondence control system
- Maintain a correspondence control system to include all incoming and outgoing correspondence
- Develop and implement a commitment control system for the project including a status system for all internal and external commitments
- Identify and comment on cross-sector project implementation issues
- Provide input on lessons learned
- Assess completed projects and programs
- Support improved project performance management by providing input and training on the design of the project performance report and the effective monitoring of key indicators
- Provide input and assistance to improve the quality of self-evaluation in programs and projects.

1.8 PROJECT ORGANIZATIONS

1.8.1 Organizational Needs

A project organization structure is a temporary structure designed to manage the project to its successful conclusion. The structure aims to facilitate the definition, coordination, and implementation of project activities. A project organization will be dynamic in size and extent, and will evolve throughout the development life cycle. However, all project organizations must establish a "core team", which will form the project management organizational backbone. The core team will be comprised of the project manager and the project leads of the major components of the project. A consistent core team facilitates the gathering and maintenance of the project's collective knowledge and subsequently enhances decision making.

The project organization seeks to establish interactions and channels of communication among the team members with an aim to reduce disruptions, overlaps, and conflict. The structure defines the relationships among members of the project management and the relationships with the external environment. The structure allows for decisions to be made at their most appropriate levels, and the organization describes the responsibilities, goals, limits of authority, relationships, skills, knowledge, and experience required for all roles in the project.

Each project has its unique characteristics, and the design of an organizational structure should consider the organizational environment, the project characteristics in which it will operate, and the level of authority the project manager is given. A project structure can take on various forms with each form having its own advantages and disadvantages (this will be discussed further in Section 1.8.3).

1.8.2 Organizational Challenges

A properly designed project organization is essential to project success. An organization chart has a pyramid form and defines the levels and scope of authority and responsibility—the higher the position, the greater is the authority. The relative locations of the positions on the organization chart designate formal and informal supervision and lines of communication between the individuals.

The project organization establishes the formal relationships among project manager, the project core-team members, the performing team members, the organization, management, beneficiaries, and other project stakeholders. This organization must facilitate an effective interaction and integration among all the major project participants and achieve open and effective communication among them.

The project structure will evolve at the different phases of the project, calling upon a different set of internal and external project players. The core-team is essential to maintain project consistency while the structure is to be designed to facilitate the interaction of people to achieve the project goals within the specified constraints of scope, time, budget, and quality. The project manager needs to find a balance between a formal and informal environment in which team members can fulfill their assignments and duties. To ease mutual collaboration, the structure must provide bilateral communication channels between individual team members, avoiding centralization through the project manager.

1.8.2.1 Factors in Designing a Project Structure

The goal of the project organization structure is the accomplishment of the project goals while providing harmony of individual efforts. The project manager's principal responsibility is to ensure that the project is organized

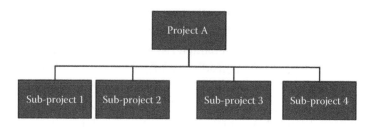

FIGURE 1.18
Basic project organization

in a way that all of the components, parts, subsystems, and organizational units fit together (Figure 1.18).

Most projects are characterized by the division of labor and task inter-dependencies, creating the need for integration to meet project objectives. The project work is organized around a hierarchical chart—the work breakdown structure (WBS) (discussed in Chapter 3). This divides the overall project goals into specific goals for each project area or component. The highest level of the WBS constitutes the structure of the project core-team. Each assigned core-team member will then develop a subordinated organizational structure. To ensure that the total efforts contribute to the overall project goals, these substructures, and the various components they handle, are integrated.

The major design factor that influences the process of developing a project management structure is the specialization and discipline required for each project area or component. This is greatest when there are many project components that have different specializations. Projects can be highly specialized and focus on a specific area of development, or have different broad specializations in many areas of development. For large projects that have multiple specializations or discipline areas, each area may have a different need, from differences in goals, approaches, and methodologies, all of which influence the way the project will implement its activities.

1.8.2.2 Corporate Organizational Structure Factors to Consider

In general, corporate organizational structures are at their most efficient when they centralize a specialization or discipline.

In such a model:

- The marketing activities promote products with the goal of staying competitive in the market.

- The purchasing function concentrates on procurement activities.
- Manufacturing focuses on rolling out the finished products.
- The sales team sells the finished product in the marketplace.
- The accounting function focuses on financial activities.
- The human resources function handles the hiring, training, and firing activities.

Other functions cover areas such as finance, legal, distribution, research, technology, and administration, to mention the major ones. Consequently, the organizational structures are functional and vertical.

Among the numerous types of organization management structures, the functional organizational structure is the most common. The functional model is structured hierarchically with a strong concept of subordination. Most companies in the modern era rely upon this functional/hierarchical model (Figure 1.19).

In the functional organization, similar function-based jobs done by the employees are grouped together in silos in the functional structure–based organization. Specialization is centralized and employees who are doing these specialized jobs constitute a unique department. Each department usually has a department head.

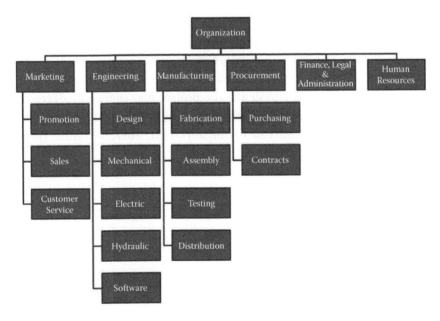

FIGURE 1.19
Functional organization

The benefits in utilizing a structure, which relies upon the functional model, are

- The chain of command is linear and sound.
- The development of professional expertise is attained by grouping specialists as a single unit.
- Focused leadership and guidance nurture the competencies and skills.
- Career paths exist for the employees to grow within the organization upwards as well as sideways in the organization.

While specialization allows each organizational structure to maximize productivity to attain departmental goals, this generates a multitude of cross-functional issues the project managers need to address in designing an effective project organization structure.

1.8.3 Types of Project Organizations

The key factor to consider when deciding on the design of the project organizational structure, especially within an existing organization, is the extent of authority and responsibility that sponsorship management is prepared to delegate to the project manager. An important function of the organizations' sponsorship management is to support an organization that fully supports project management. This is achieved by refining the organization to emphasize the nature of the projects and adapting how roles and responsibilities are assigned.

The organization needs to define the project manager's job, degree of authority and autonomy, and relationship to both the organization, other projects, and to other units in the organization. There are three major types of project organizations:

1. Functional
2. Project-based (projectized)
3. Matrix

1.8.3.1 Functional Organization

This structure is by far the most utilized of the organizational methods. This method performs best when used for routine work functions and

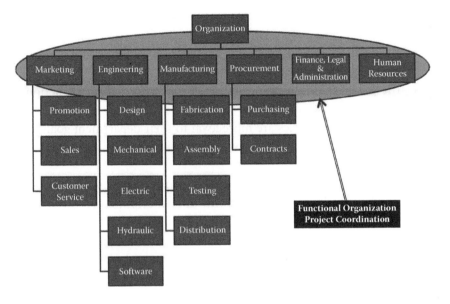

FIGURE 1.20
Functional project organization

the upholding of quality and work standards. Functional organization structures assign projects in two different ways (Figure 1.20).

One way involves the project being assigned to a specific functional manager who then coordinates with the other departments for each of them to contribute. Alternatively, projects can be shuffled around to different departments where each department manager ensures that their parts of the work have been completed.

One major disadvantage of functional organization is that the department may not have all of the specialists needed to work on a project.

When cross-functional project management is necessary, portions of the project are performed within separate departmental units. Retrieving information or efforts from other departments is done by requesting assistance from the other unit by routing the request through the head of the unit to the head of the other unit.

Another disadvantage is that personnel assigned to project activities may have other responsibilities in the department that could impact their ability to meet project deadlines. Subject matter experts may also be assigned to other projects.

The drawbacks of the functional organizational structure are

- The decision-making process is bureaucratic and the speed of resolving problems is slow and inefficient.
- The flow of communication and synchronization between functional departments is complicated.
- Functional employees are more loyal to their department goals and lack the broader view of the overall project objective.

In conclusion, this type of project organizational structure fails when used in facilitating complex and multidiscipline projects. There is very little individual accountability for any project management tasks that need to be performed, and this organizational structure often lacks employee recognition, measurement, and reward for project performance.

1.8.3.2 Project Based (Projectized)

The project manager in this structure has total authority over the project and can acquire the resources needed to accomplish project objectives from within or outside the organization, subject only to the scope, quality, and cost constraints identified as project targets (Figure 1.21).

In this structure, personnel are specifically assigned to the project and report directly to the project manager. The project manager assumes responsibility for the performance appraisal and career progression of all project team members while on the project.

Complete line authority over project efforts leads to rapid reaction time and improved responsiveness. Moreover, project personnel are retained on an exclusive rather than shared or part-time basis. Project teams develop a strong sense of project identification and ownership, with deep loyalty efforts to the project and a good understanding of the nature of the project's activities, mission, and goals.

Pure project-based organizations are more common among large projects and those requiring multiple disciplines. The most obvious advantage of project-based organizations is that there is no need to negotiate with other departments for resources. Other advantages of this organization are that the team members are usually familiar with each other, and they bring applicable knowledge of the project.

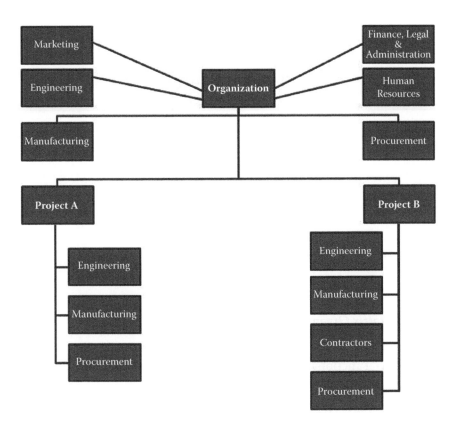

FIGURE 1.21
Project-based organization

Once the project is completed, this structure disbands. This raises concerns on the reallocation of people and other resources when projects are completed.

One major disadvantage may be inefficient use of personnel, as project team members are generally dedicated to only one project over the whole life cycle of the project. It is rare that any resource will be utilized to its maximum availability. This also raises a cost concern, compounded by a duplication of resources across the different projects, as each would find difficulty in employing fully its dedicated resources.

A criticism of this structure is that it may be inefficient in transferring technology and the use of resources between projects, unless a continuous lessons-learned and peer exchange mechanism is in place. Also, and especially for short duration projects, by the time the members actually begin performing as a cohesive team, the project is completed and the organization disbanded.

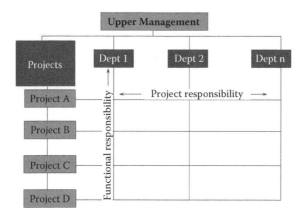

FIGURE 1.22
Matrix organization

1.8.3.3 Matrix Organization

Matrix organization is a project management structure that evolved from the functional organization and project-based structures (Figure 1.22). This structure aims to combine the best components of the two other structures. This model functions very well when there are multiple projects being coordinated concurrently. Project managers are assigned from the functional units. However, it is the functional managers who oversee the staffing, training, job assignment, and evaluation of the project's personnel. The functional specialists are assigned to one or more projects and manage their activities to achieve their objectives. Project managers have more of a coordinating role. However, all functional managers must give priority to projects rather than their departmental objectives.

Matrix organizations allow departmental units to focus on their specific technical competencies and expand these, and allow projects to be staffed with specialists from throughout the organization. For instance, subject matter specialists may report to one department but would be assigned to various project managers from other departments.

The major advantages of matrix management are

- Efficient allocation of all resources in the organization, especially scarce specialty skills that cannot be fully utilized by one project

- Allows team members to share information more rapidly across the unit boundaries
- Easier for a department manager to assign an employee to another manager without making the change permanent
- Easier to accomplish work objectives in an environment when task loads are shifting rapidly between departments

The main disadvantage of matrix organizations is that the reporting relationships are complex. Some project members might report to department managers, for whom little work is done, while actually working for one or more project managers. Individual employees will report to at least two managers, which can often lead to conflicts of assignment priorities. These problems can be avoided through good communication and solid leadership between managers. Project members need to develop strong time management skills to ensure that they fulfill the work expectations of multiple managers.

Matrix management places some difficulty on the project managers because they must work closely with other managers and employees in order to complete the project. This organization requires communication and cooperation between multiple department managers and project managers that all need to share the limited availability from the same resources. As the department managers may have different goals, objectives, and priorities than the project managers, these would have to be addressed at the earliest assignment opportunity.

An approach to help solve assignment conflicts is a variation of the matrix organization, which includes a coordinating role that either supervises or provides support to project managers. In some organizations, this is the project management office (PMO), dedicated to provide expertise, best practices, training, methodologies, and guidance to project managers.

PRACTICE EXERCISE

Reflect on projects that are currently being undertaken (or have recently been completed) in your organization. What kind of organization structure was in place? What do you think were the advantages and disadvantages of this structure in your organization?

1.9 PMI BODY OF KNOWLEDGE

1.9.1 PMI: Nine Knowledge Areas

The Project Management Institute publishes the *Project Management Body of Knowledge* (PMBOK®), which describes the generally accepted knowledge and practices applicable to most projects most of the time, and that have widespread consensus about their value and usefulness.

The nine project management knowledge areas describe the key competencies that project managers must develop.

Project integration management is an overarching function that affects and is affected by all of the other knowledge areas, which are grouped into four core knowledge areas and four facilitating areas.

The four core knowledge areas of project management include project scope management, project time management, project cost management, and project quality management.

These are core knowledge areas because they lead to specific project objectives. Brief descriptions of each core knowledge area are as follows:

- **Project scope management** involves defining and managing all the work required to complete the project successfully.
- **Project time management** includes estimating the duration to complete the work, developing a project schedule, and ensuring timely completion of the project.
- **Project cost management** describes the processes involved in preparing and managing the budget for the project.
- **Project quality management** ensures that the project will satisfy the undertaken quality requirements.

The four facilitating areas are project human resource management, project communications management, project risk management, and project procurement management.

These are called facilitating areas because they are the processes through which the project objectives are achieved. Brief descriptions of each facilitating knowledge area are as follows:

- **Project human resource management** is concerned with planning, acquisition, and effective management of the project team.

- **Project communications management** involves generating, collecting, disseminating, and storing project information.
- **Project risk management** includes identifying, analyzing, responding to, and controlling risks related to the project.
- **Project procurement management** involves acquiring or procuring products, goods, and services for a project from outside the performing organization.

KNOWLEDGE AREA	KEY TOOLS AND TECHNIQUES
Integration management	Project selection criteria, project management methodology, stakeholder analysis, project charters, project management plans, change control and configuration management, project review meetings, project management software
Scope management	Project scope statements, requirements analysis, work breakdown structures, statements of work, scope management plan, scope change control
Time management	Activity definitions, duration estimating techniques, project network diagrams, critical path analysis, critical chain scheduling, Gantt charts, resource leveling, crashing, fast tracking
Cost management	Activity cost estimating techniques, net present value and payback analysis, reserve analysis, earned value management, cost management plan, financial software
Quality management	Cost-benefit analysis, quality control charts, Pareto diagrams, Ishikawa (fishbone) diagrams, quality audits, statistical methods
Human resource management	Organizational charts, interpersonal skills, motivation techniques, communication techniques, team building and contracts, roles and responsibilities matrices, resource histograms, resource leveling, conflict management techniques

KNOWLEDGE AREA	KEY TOOLS AND TECHNIQUES
Communications management	Stakeholder identification, communications management plan, communications media methods, communications infrastructure, status reports, conflict management
Risk management	Risk management plan, risk identification, probability/impact analysis, risk ranking, Monte Carlo simulation, variance/trend analysis, risk reporting
Procurement management	Make-or-buy analysis, procurement contracts, requests for proposals or quotes, source selection, negotiating, procurement performance management, payment systems

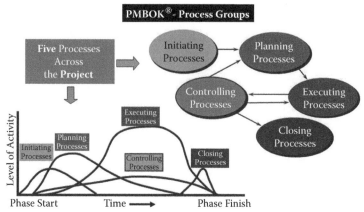

PMBOK is a trademark of the Project Management Institute Inc., which is registered in the United States and other nations.

FIGURE 1.23
PMI—process groups

1.9.2 PMI: Process Groups

The five project management process groups described in Figure 1.23 are initiating, planning, executing, monitoring and controlling, and closing.

Initiating	Defines and authorizes the project or a project phase
Planning	Defines and refines objectives and plans the course of action required to attain the objectives and scope that the project was undertaken to achieve
Executing	Defines the processes performed to complete the work defined in the project management plan
Monitoring and controlling	Defines the processes to monitor and evaluate progress, and to respond to variances to the project management plan and to apply corrective action when necessary to meet project objectives
Closing	Finalizes and formalizes acceptance of the project or a project phase

1.9.3 Mapping Processes to Knowledge Areas

The level of interaction of the five process groups emphasizes their strong relational dependence. The processes are neither sequential nor chronological. The interrelationship is a function of progressive elaboration, where details are developed along the project phases.

Mapping Processes to Knowledge Areas					
	Initiating Processes	Planning Processes	Executing Processes	Controlling Processes	Closing Processes
Integration		4.1 Project plan development	4.2 Project plan execution	4.3 Integrated change control	
Scope	5.1 Initiation	5.2 Scope planning 5.3 Scope Definition		5.4 Scope verification 5.5 Scope change control	
Time		6.1 Activity definition 6.2 Activity sequencing 6.3 Activity duration estimating 6.4 Schedule development		6.5 Schedule control	
Cost		7.1 Resource planning 7.2 Cost estimating 7.3 Cost budgeting		7.4 Cost control	
Quality		8.1 Quality planning	8.2 Quality assurance	8.3 Quality control	
Human Resources		9.1 Organizational planning 9.2 Staff acquisition	9.3 Team development		
Communication		10.1 Communications planning	10.2 Information distribution	10.3 Performance reporting	10.4 Administrative close out
Risk		11.1 Risk management planning 11.2 Risk identification 11.3 Qualitative risk analysis 11.4 Quantitative risk analysis 11.5 Risk response planning		11.6 Risk monitoring and control	
Procurement		12.1 Procurement planning 12.2 Solicitation planning	12.3 Solicitation 12.4 Secure selection 12.5 Contract administration		12.6 Contract close out

PMBOK is a trademark of the Project Management Institute Inc., which is registered in the United State and other nations.

FIGURE 1.24
PMI—knowledge areas

Active and proactive project management is required throughout the duration of the project. Thus it must be continually planned, monitored, and controlled.

Project management process groups are not project phases. The process groups are present for each phase. Process groups and project phases are interdependent and require integration.

From the point when objectives are developed, problem and solution analyses are performed and completed. This may lead to a change of constraints. During the project execution, risks are mitigated and contingencies are applied. Changes are managed and resources are realigned, as unforeseeable or unpreventable circumstances are addressed (Figure 1.24).

2

Project Initiation

2.1 CHAPTER OVERVIEW

This chapter covers the origination and initiation phases of a project. Special attention and details are given to the process of project selection, and how the project fits in the corporate strategic alignment.

The central portions of this chapter present project portfolio management, stakeholder management, scope management, and the key project documents, which define the foundations of the project plan.

Techniques for establishing the project organizational structure, as previously described in Chapter 1, are reviewed, as well as the relationship roles of the project manager with project players.

2.2 PROJECT ORIGINATION

2.2.1 Origins of Projects

The challenge of a performing organization is to sustain and maintain current operations while addressing both growth and opportunities.

Simply stated, the organization continuously asks the question "Are we doing it *right*?"

- Right market
- Right products
- Right organization
- Right skills
- Right investments
- Right size

The performing organization will seek out the operational inhibitors such as:

- Waste
- Obsolescence
- Heavy process points
- Costly steps
- Low quality
- Inadequate response

The performing organization will also assess the enablers and opportunities such as:

- New market
- New product
- Alliances

These reasons for change, and thus projects, contribute to the decision-making process by the performing organization management. Some of these reasons may be forced upon the organization because of internal factors, such as inefficiency or obsolescence, or external factors, such as competitive forces or statutory requirements, environmental regulations, and legislative restrictions. This places the organization in a reactive mode. Ideally, the organization wants to be in a proactive state.

Project origination is the one phase in the project management life cycle for which the performing organization management is solely responsible: The project manager will not be assigned to the project until the next phase—initiation. Nevertheless, the project manager should understand why the project is being launched.

The performing organization management will identify and assign project sponsorship at this stage.

The purpose of project origination is to provide a formal mechanism for recognizing and identifying potential projects within the organization, to evaluate projects proposed for the next cycle, and to reach a decision on projects to be selected. During this phase, a project proposal is developed to create a product or develop a service that can solve a problem or address a need in the performing organization. The proposal's business case is reviewed and validated, and the choice of the proposed solution is assessed. Foremost, the proposed project is gauged for its consistency with the organization strategic and operational plans. It is also appraised for the value it will create, and whether it is within the investment and budget guidelines.

A formal project origination process is important because it creates a focal point for assessment and analysis of all proposed initiatives, directs resources to the right projects, and focuses the organization on projects that will deliver the greatest value. Each organization has its own project approval process. However, this process revolves around three major steps:

1. **Develop project proposal**—the business case is established and the initial project boundary and parameters are defined.
2. **Evaluate project proposals**—quantitative financial analysis is performed, such as cost/benefit analysis and ROI, and proposed projects are evaluated against a set of pre-established strategic and operational business criteria.
3. **Select projects**—following agreement on the project's feasibility, this is ranked with other proposed projects, and a decision is formally made regarding the project proposal.

To make a meaningful evaluation, the decision-making body must possess sufficient information on the project's business case and the viability of its proposed solution. The competing projects must be evaluated and compared using a consistently applied methodology, and the selection process must consider the project's fit with the organizational strategic/operational plans.

Once a project is selected, funding and/or further management commitment is required to progress to project initiation. Formal project sponsorship is then formally established.

The project proposal process in the origination phase may actually be part of the total budget cycle, serving as the justification for funding requests. In this case, the project proposal is to include a budget estimate for the total cycle and a funding request to proceed to and perform the project's initiation phase.

PRACTICE EXERCISE

Consider the projects that are currently being undertaken (or have recently been completed) in your organization. Identify the principal driver(s) of these projects and whether they exist to address operational inhibitors or promote opportunities.

2.2.2 Enabling Documents

A proposal for a project may be instigated from anywhere in the performing organization; however, a project sponsor must be assigned to ensure that the key proposal documents are established. As seen above, the sponsor would steer the project through evaluation and selection. A proposal team is often established to develop the business case and proposed solution documents, which describe the product of the project, the benefit to the organization, alignment with the organization's strategic and operational plans, and a high-level estimate of the required resources and costs. On smaller projects, the project sponsor may perform this task without the assistance of a team.

As stated above, to make a meaningful evaluation, the decision-making body must possess a thorough project proposal. The key enabling document is the business case, which provides detailed information on the project's drivers and the viability of its proposed solution.

2.2.2.1 The Business Case

As previously explained in Chapter 1, the business case is a mandatory step prior to any project launch. Business cases must clearly state the business drivers and reasons for the proposed project.

The business case (seen in Figure 2.1) usually consists of two major facets: the current situation, which states the need, issue and/or problem, and the solution proposal, which presents the desired business outcomes and benefits, and the roadmap to meet these.

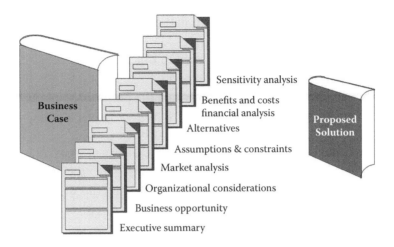

FIGURE 2.1
Business case contents

There is no hard and fast rule for the structure of the business case; however, all business cases must include the following:

1. The executive summary highlights the key points in the business case. It states the business problem by providing a thorough and objective presentation of a business issue that requires a timely solution. The summary must highlight the benefits and the return on investment. It must also demonstrate how the business case responds to the corporate strategic and/or operational alignment.
2. The business opportunity describes the incentive for the project that the business case will describe and propose. The business opportunity includes a definition, a statement of scope, and a discussion of objectives that the project will help the organization achieve. It will explain how the expected outcome of the project supports and aligns to the organization's strategic or operational intent.
3. The organizational considerations examine the current organization and highlight the areas where the proposed project will affect the future structure.
4. The market analysis examines changes in the business environment, such as clients, competitors, suppliers, industry standards, and legislation.
5. The constraints are internal and external. They will cover the current and future constraints originating from schedule, resource, budget,

staffing, technical, and other limitations that may affect the success of a project.

6. The alternatives section analyzes the different choices that can be considered and their respective merits.

7. The assumptions are events on which the proposed solution of the business case is based. Assumptions must be validated for a project to succeed.

8. The benefits and costs provide a comprehensive assessment of both expected benefits and anticipated costs throughout the total product life cycle. These benefits are to be qualified in terms of their impact on the organization's operational performance.

9. The financial analysis compares benefits to costs and analyzes the value of a project as an investment. The analysis must be credible on the return on investment and payback period, by identifying the ownership costs required for the project development and the operational costs for the utilization of the end-product or service. Projects of an anticipated duration greater than one year must also consider net present value and internal rate of return. The financial analysis may also need to include a cash flow statement.

10. The sensitivity analysis evaluates the project risks and presents the mitigation and contingency plans, and the proposed solution's risk exposure.

11. The proposed solution should offer an approach to solving the business issue/problem and demonstrate that the project is both viable and beneficial. The proposed solution must emphasize the merits it has over the alternative solutions that exist. The proposed solution should address only the agenda described by the business problem, and the approach narrative should tie the solution to the existing business problem in cost/benefit terms.

12. Project objectives should be explicitly stated and be aligned with the organization's strategic and operational plans and infrastructure. The project key stakeholders must be referenced, as well as the nature of sponsorship and upper-management support and commitment.

13. An organizational impact analysis should describe the anticipated impact of the project on the organization's people, processes, and technology.

14. Budget/resources estimates (mostly in order of magnitude) should consider all cost components of the ownership and operations cycle, including support, maintenance, and other recurring costs.

15. A high-level project implementation plan is to be established spanning the initiation to close-out phases. Any plan at this stage can only be tentative and must be considered as preliminary. The subsequent initiation and planning phases will refine the plan. The proposed project development plan will include:

- Deliverables schedule
- Phase/stage definitions
- Macro activities
- Workload estimate/breakdown
- Project schedule
- Required resources
- Funding requirements
- Project leadership team, project governance team
- Project controls and reporting processes

16. The recommendations summarize the main points of a business case and offer suggestions on how to proceed with the project.

2.3 PROJECT SELECTION

2.3.1 Projects and Project Portfolio Management

Projects contend with each other for limited and constrained funding and resources. It is essential for any organization to establish a solid and pragmatic process for selecting those projects that bring the most value and reap the best benefits.

Each organization will choose how best to approach the management of projects and the level of discipline it wishes to follow to select projects that will fulfill the strategic and operational goals. A haphazard and/or subjective selection technique often leads to missed opportunities and a waste of funding and resources. An objective and informed project selection process does not guarantee total success; however, it will improve the chances of success.

Whatever selection process is followed, a mechanism must exist that centralizes and brings together all project contenders, to be assessed and compared, and from which the performing organization can determine the priorities to allocate to each project. This mechanism has many names; however, the most applicable and efficient is a project portfolio.

PMI offers a very comprehensive definition.

"A project portfolio is a collection of projects and/or programs and other work that is grouped together to facilitate the effective management of that work to meet strategic business objectives. The projects or programs (hereafter referred to as "components") of the portfolio may be mutually independent or directly related. At any given moment, the portfolio represents a "snapshot" of its selected components that both reflect and affect the strategic goals of the organization—that is to say, the portfolio represents the organization's set of active programs, projects, subportfolios, and other work at a specific point in time." (PMBOK® Guide, p. 8)

Project portfolio management (PPM) groups programs/projects so they can be managed as a portfolio. PPM ensures that programs/projects and expenditures are aligned with corporate strategy and operational objectives (Figure 2.2).

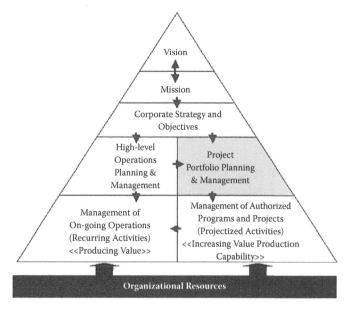

FIGURE 2.2
Portfolio—PMI pyramid

Project portfolio management (PPM) is a framework that translates strategy into programs/projects and aligns these to the financial and capacity management disciplines of the company. To be fully effective, PPM is best extended to include those initiatives generated from operations.

The extent of PPM is scalable and is to be tailored to the organization's environment. Not all programs/projects need to be managed by a PPM, and thresholds can be set such as below a certain budget limit or less than a certain schedule duration. Similarly, operational managers may consider that certain routine support and/or maintenance projects are best managed outside of a PPM environment.

A key benefit of PPM is that it provides executives a synthetic view of how programs/projects contribute to the organization's strategic intent and fulfill operational objectives (Figure 2.3). It also assists executives to assess where funding/resources are needed for program/project contenders.

Project portfolio management will provide for the periodic review, direction, and allocation of priorities and resources across the portfolio. This will take into account:

- The business unit or organization's strategy and objectives
- Changes in internal or external environment
- Business operational performance
- The status, expected benefits, and risks of all portfolio programs/projects

PPM allows for the efficient allocation of internal/external resources for those programs/projects currently in the portfolio, or which are to be included. Each program/project is subsequently assigned a contribution ranking to the portfolio's strategic intent.

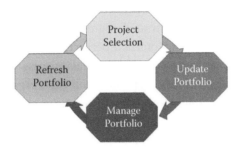

FIGURE 2.3
Portfolio cycle

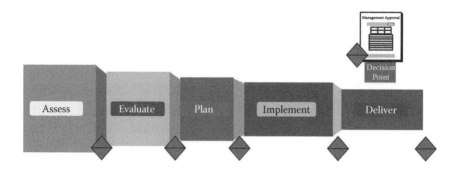

FIGURE 2.4
Project funnel process

The PPM is the overarching management and governance process for the identification, evaluation and selection, prioritization, authorization, and performance review of programs/projects within the portfolio, and their alignment to the strategic intent and operational goals.

Programs/projects in the portfolio are subordinated to decisions based on their alignment with corporate strategy, viability, portfolio resource availability, priorities, and the evolution of the portfolio contents (see Figure 2.4).

Project portfolio management ensures that the portfolio stays aligned with business objectives. This involves following a continuous process by which programs/projects are evaluated, prioritized, selected, and managed at formal points such as a portfolio gate reviews.

2.3.2 Aligning the Project to the Organizational Strategy

An organization's strategic intent can only be accomplished by an efficient and focused approach of management by projects. The strategic intent is translated into road maps constituting sets of programs/projects, all requiring allocation of resources. Programs/projects are periodically reviewed to confirm their continued alignment with objectives and their priority is adjusted depending on their performance.

Project portfolio management is a pivotal process for the successful fulfillment of strategic intents. Depending on each organization, the visibility given to the portfolio by all levels of executive and operational management will be a key critical success factor. If selection decisions are not made at the portfolio level, by default the project portfolio is the end result of individual project choices made one at a time with little regard for the impact that one project has on the next.

Members of the portfolio executive group or steering committee, sponsors, and key stakeholders of all programs/projects in the portfolio are to receive pertinent and up-to-date progress/status information on the portfolio performance. Operational management need to be informed as to the organizational readiness requirements they have to prepare to fulfill the business benefits after completion/handover of any subset of a program/project.

Upper executive management are to be informed on the investment/funding requirements for the medium and long terms. In many cases, this is done by a "rolling" quarterly financial statement covering three to five years, or more depending on the organization.

Upper and executive management must inform the project portfolio management about any anticipated changes in the company's direction, to ensure the alignment of programs/projects to the modified strategic intent(s).

PRACTICE EXERCISE

Does project portfolio management exist in your organization? If so, can you describe its process? If you do not have such a process, do you think it is necessary to introduce one? How would you promote this?

2.3.3 Selection and Prioritization of Projects

The assessment, prioritization, and selection of projects are best done by a project selection committee (see Figure 2.5). This will certainly be the case when functioning with a PPM system. Outside of this, the informal decision making should seek inspiration from the PPM system to lend credibility to the process.

The project sponsor should gain an understanding of the organization's formal and informal project selection processes. Being knowledgeable about these processes, providing all pertinent project information to the organization's project selection committee, and introducing the proposal

FIGURE 2.5
Proposal evaluation

to the committee at the appropriate time will improve the project's chances of being selected.

In order to implement an effective project selection and priority process, the project selection committee roles and responsibilities should be clear:

- Evaluating project proposals on the basis of the selection criteria
- Accepting or rejecting proposals
- Publishing the score of each proposal and ensuring the process is open and transparent
- Balancing the portfolio of projects for the organization
- Evaluating the progress of the projects in the portfolio
- Reassessing organizational goals and priorities if conditions change

Initially, the committee can filter out projects that do not at least meet the following:

- Fit in with the organization's strategic plans
- Fit into the existing (or projected) organizational processes
- Compliant with the organization's standards
- Conform to available funding allocations and limits

The contents of the full project proposals assessment criteria will vary between organizations.

At a minimum, the project selection committee evaluates how the project proposal:

- Aligns to strategic goals
- Aligns to core competencies
- Responds to stated business issue/problem and goals
- Presents expected outputs and outcomes that are consistent with the goals described
- Demonstrates clearly how the goals will be fulfilled—project plans, schedules, resources
- Includes indicators to monitor implementation progress and performance (intermediary outputs and milestones) towards achieving their final outputs and goals
- Manages risks
- Defines the organizational impacts
- Presents operational sustainability
- Describes clearly and realistically the funding requirements
- Demonstrates benefits and value
- Complies with national or local laws and regulations

Management weights each criterion by its relative contribution and importance to the organization's goals and strategic plan. The project selection committee evaluates each project proposal by its relative contribution or benefit to the selection criteria. The committee assigns a spectrum of values for each criterion ranging from low (0) to high (10). This value represents the proposal's fit to the specific criterion. The aggregate of all assigned values determines the ranking of the project proposal.

The project sponsor follows the selection process, attending the project selection committee meetings as required. The committee determines which projects get approval to proceed to project initiation, which project proposals require more information for further evaluation, and whether some projects should be removed from further consideration. Committee decisions are documented and communicated to the project sponsor.

The main measurement of success for project selection is the consensus of the performing organization management that the proposed projects were weighed fairly, and that the ones with a compelling business case received approval.

2.3.4 Quantitative and Qualitative Methods

Project proposals have to demonstrate viable and realistic financial and economic data for the total costs of ownership and operations. This is applicable only for projects that produce a quantitative tangible benefit that can be measured in monetary terms. Many projects that will enhance operational efficiency, increase organizational skills, increase customer satisfaction, or participate indirectly to the benefits of other projects will be hard-put to calculate the financial effects. Project managers must be conversant with the following key quantitative techniques:

- Benefit-cost ratio (BCR)
- Internal rate of return
- Payback period
- Return on investment

The backdrop to these techniques is illustrated in Figure 2.6.

A major consideration in BCR, and the other techniques presented here, is the notion of the time value of money. Present value and net present value are used to calculate this.

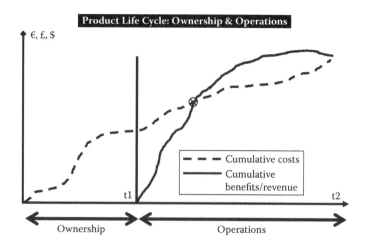

FIGURE 2.6
Product life cycle—ownership and operations

2.3.4.1 Present Value

Present value is the value on a given date of a future payment or series of future payments, discounted to reflect the time value of money and other factors such as investment risk. Present value calculations are widely used in business and economics to provide a means to compare cash flows at different times on a meaningful "like to like" basis.

The most commonly applied model of the time value of money is compound interest. To someone who can lend or borrow for t years at an interest rate i per year (where interest of 5 percent is expressed fully as 0.05), the present value PV of the receiving C monetary units t years in the future is:

$$C_t = C (1 + i)^{-t} = C / (1 + i)^t$$

$$PV = C / (1 + i)^t$$

Example

If the current value C is $1 m, the interest rate i is 10 percent per year, expressed as 0.10, and t is 3 years,

$PV = \$1m / (1 + 0.10)^3$
$PV = \$1m / (1.331)$
$PV = \$751\ 315$

2.3.4.2 Net Present Value

Net present value (NPV) is the total present value (PV) of a time series of cash flows (also called discounted cash flow or DCF). It is a standard method for using the time value of money to appraise long-term projects. All future estimated input and output cash flows are discounted to give their present values. It measures the excess or shortfall of cash flows, in present value terms.

$$\text{NPV} = PV \text{ input cash flow} - PV \text{ output cash flow}$$

Example

Project A and Project B are competing for approval. Project A shows a Benefit of 14,000 and a Cost of 9,000. Project B shows a Benefit of 15,000 and a Cost of 10,000. Both have a +5,000 net result before applying *PV.*

The table shows that on applying *PV* and determining NPV over the same time period of 5 years, Project B is more profitable.

Interest 10%	0.10					
$(1 + i)^t$	1.10000	1.21000	1.33100	1.46410	1.61051	1.77156

Project A	Year 1	Year 2	Year 3	Year 4	Year 5	Total
Benefits	0	2,000.00	3,000.00	4,000.00	5,000.00	14,000.00
PV Input	0.00	1,652.89	2,253.94	2,732.05	3,104.61	9,743.50
Costs	−5,000.00	−1,000.00	−1,000.00	−1,000.00	−1,000.00	−9,000.00
PV Output	−4,545.45	−826.45	−751.31	−683.01	−620.92	−7,427.15
NPV	−4,545.45	826.45	1,502.63	2,049.04	2,483.69	2,316.35

Project B	Year 1	Year 2	Year 3	Year 4	Year 5	Total
Benefits	1,000.00	2,000.00	4,000.00	4,000.00	4,000.00	15,000.00
PV Input	909.09	1,652.89	3,005.26	2,732.05	2,483.69	10,782.98
Costs	−2,000.00	−2,000.00	−2,000.00	−2,000.00	−2,000.00	−10,000.00
PV Output	−1,818.18	−1,652.89	−1,502.63	−1,366.03	−1,241.84	−7,581.57
NPV	**−909.09**	**0.00**	**1502.63**	**1366.03**	**1241.84**	**3201.41**

2.3.4.3 Benefit-Cost Ratio (BCR)

This is the ratio to identify the relationship between the total life cycle cost and benefits of a proposed project. Cost covers both ownership and operational costs, whereas benefits are usually recorded in operations. It is important to define the start point of the cost accrual. The BCR is calculated at a given dateline during operations. This dateline defines at what stage the cumulative costs and benefits can compared (see Figure 2.7).

The BCR is a simple and rapid way to assess the financial viability of a project proposal.

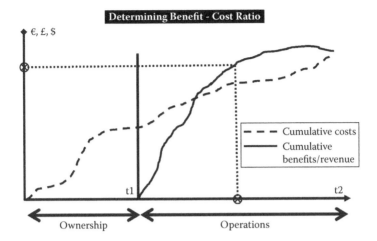

FIGURE 2.7
Determining benefit-cost ratio

Example 1

The estimated ownership cost is $1.5 m and the estimated operational costs until the chosen dateline is $25.5 m, giving a cumulative cost of $27 m. The estimated benefits in operations at the dateline are $36 m. The BCR is $36 m / $27 m = 1.33

Example 2

The estimated ownership cost is $12 m and the estimated operational costs until the chosen dateline is $51 m, giving a cumulative cost of $63 m. The estimated benefits in operations at the dateline are $49 m.
   The BCR is $63 m / $49 m = 0.78
BCR is viable when equal to or greater than one (1). In the project proposal selection process, BCR can be used for ranking.

PRACTICE EXERCISE

Which of the following two projects is the most beneficial for the organization?

| Interest 10% | 0.10 | | | | | |
| $(1 + i)^t$ | 1.10000 | 1.21000 | 1.33100 | 1.46410 | 1.61051 | |

Project A	Year 1	Year 2	Year 3	Year 4	Year 5	Total
Benefits	0	60,000.00	300,000.00	400,000.00	450,000.00	1,210,000.00
Costs	−100,000.00	−80,000.00	−200,000.00	−240,000.00	−250,000.00	−870,000.00
Benefits-Costs	−100,000.00	−20,000.00	100,000.00	160,000.00	200,000.00	340,000.00
NPV	0.00	0.00	0.00	0.00	0.00	0.00

Project B	Year 1	Year 2	Year 3	Year 4	Year 5	Total
Benefits	50,000.00	250,000.00	450,000.00	500,000.00	450,000.00	1,700,000.00
Costs	−300,000.00	−200,000.00	−250,000.00	−300,000.00	−300,000.00	−1,350,000.00
Benefits-Costs	−250,000.00	50,000.00	200,000.00	200,000.00	150,000.00	350,000.00
NPV	0.00	0.00	0.00	0.00	0.00	0.00

2.3.4.4 Internal Rate of Return (IRR)

The internal rate of return (IRR) is the interest rate that makes the net present value of all cash flows from a particular project equal to zero. The higher a project's internal rate of return, the more desirable it is to undertake the project. As such, IRR can be used to rank several proposed projects an organization is considering. Assuming all other factors are equal among the various projects, the project with the highest IRR would probably be considered the best and undertaken first.

The IRR is determined by numerical iterations. This is cumbersome and using a financial calculator would simplify this processes.

2.3.4.5 Payback Period

The payback period in projects refers to the period of time required for the return on an investment to "repay" the sum of the original investment. Graphically, it is the point where both the cumulative cost and benefits curves cross. This point is the break-even point. Shorter payback periods are obviously preferable to longer payback periods (see Figure 2.8).

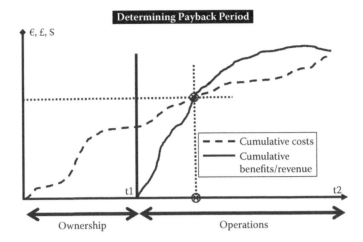

FIGURE 2.8
Determining payback period

Payback period as a tool of analysis is often used because it is easy to apply and easy to understand. When used carefully or to compare similar investments for competing project proposals, it can be quite useful. However, it has serious limitations and qualifications for its use, because, even though the time value of money can be adjusted with present value, it does not properly account for risk, financing, or other important considerations such as the opportunity cost. An implicit assumption in the use of payback period is that returns to the investment continue after the payback period.

In the table example, the payback period (break-even point) will occur during year 3.

Project B	Year 1	Year 2	Year 3	Year 4	Year 5	Total
Benefits	1,000.00	2,000.00	4,000.00	4,000.00	4,000.00	15,000.00
PV Input	909.09	1,652.89	3,005.26	2,732.05	2,483.69	10,782.98
Cumulative Benefits	*909.09*	*2,561.98*	*5,567.24*	*8,299.30*	*10,782.98*	
Costs	−2,000.00	−2,000.00	−2,000.00	−2,000.00	−2,000.00	−10,000.00
PV Output	−1,818.18	−1,652.89	−1,502.63	−1,366.03	−1,241.84	−7,581.57
Cumulative Costs	*−1,818.18*	*−3,471.07*	*−4,973.70*	*−6,339.73*	*−7,581.57*	
NPV	**−909.09**	**0.00**	**1502.63**	**1366.03**	**1241.84**	**3201.41**

PRACTICE EXERCISE

Determine the year of the break-even point for both projects.

Interest 10%	0.10					
$(1 + i)^t$	1.10000	1.21000	1.33100	1.46410	1.61051	

Project A	Year 1	Year 2	Year 3	Year 4	Year 5	Total
Benefits	0	60,000.00	300,000.00	400,000.00	450,000.00	1,210,000.00
Costs	−100,000.00	−80,000.00	−200,000.00	−240,000.00	−250,000.00	−870,000.00
Benefits- Costs	−100,000.00	−20,000.00	100,000.00	160,000.00	200,000.00	340,000.00
NPV	0.00	0.00	0.00	0.00	0.00	0.00

Project B	Year 1	Year 2	Year 3	Year 4	Year 5	Total
Benefits	50,000.00	250,000.00	450,000.00	500,000.00	450,000.00	1,700,000.00
Costs	−300,000.00	−200,000.00	−250,000.00	−300,000.00	−300,000.00	−1,350,000.00
Benefits- Costs	−250,000.00	50,000.00	200,000.00	200,000.00	150,000.00	350,000.00
NPV	0.00	0.00	0.00	0.00	0.00	0.00

2.3.4.6 *Return on Investment (ROI)*

Return on investment (ROI) is a performance measure used to evaluate the efficiency of a project investment or to compare the efficiency of a number of different project investments, in a given time-frame. To calculate ROI, first a time-frame is established, then the cumulative benefit of an investment is divided by the cumulative cost of the investment. The result is expressed as a percentage or a ratio.

ROI = (cumulative benefit - cumulative cost) / cumulative cost

ROI is a very popular metric because of its versatility and simplicity. That is, if an investment does not have a positive ROI, or if there are other opportunities with a higher ROI, then the investment should not be undertaken.

In the broadest sense, ROI measures the profitability of an investment. As such, there is no one "right" calculation. The calculation can be modified to suit the situation, as it depends on what is included as benefits and costs. A financial analyst and a product marketer may compare two same projects using different parameters. While the analyst may consider the total product life cycle costs, the marketer may only consider the operational costs.

In the example here, Project A has a project development period of one year, and operational use starts at year 2. The financial analyst will consider costs to accrue from year 1, whereas the marketer will start accrual from year 2.

Project A	Year 1	Year 2	Year 3	Year 4	Year 5	Total
Benefits	0	2,000.00	3,000.00	4,000.00	5,000.00	14,000.00
PV Input	*0.00*	*1,652.89*	*2,253.94*	*2,732.05*	*3,104.61*	*9,743.50*
Costs	−5,000.00	−1,000.00	−1,000.00	−1,000.00	−1,000.00	−9,000.00
PV Output	*−4,545.45*	*−826.45*	*−751.31*	*−683.01*	*−620.92*	*−7,427.15*
NPV	**−4,545.45**	**826.45**	**1,502.63**	**2,049.04**	**2,483.69**	**2,316.35**

2.4 PROJECT INITIATION

2.4.1 Approval to Proceed

Project initiation is a formal step in the whole management of project process. Initiation is the process of formally authorizing a new project or authorizing an existing project to continue into its next phase. The purpose of the project initiation phase is to verify the assumptions and projections made in the project origination that led to project selection and approval. During this phase of the project, the overall project parameters are reviewed and refined and key documents are developed. The initiation phase may need to complete the needs assessment, a feasibility study, or perform another analysis that was itself separately initiated.

Following approval, the project manager is formally assigned. The project manager will then work closely with the project sponsor to identify the core team members and other resources needed to develop the project charter and to define further develop the project scope, cost, and schedule. An initial project plan is produced at a level of detail sufficient to identify any additional resources needed to progress to the next phase—project planning. At the conclusion of project initiation, a decision is made either to halt the project or to proceed.

From the project manager's viewpoint, the project initiation phase is a key step, as the project manager may have had little or no input during the origination phase. The project initiation is the opportunity for the project manager to review all the enabling documents provided by the origination phase and to appreciate and understand the quantitative and qualitative methods that senior management may use to select projects.

The performing organization management plays a vital role during project initiation. The project initiation phase cannot begin until it has a project sponsor, who is the liaison between the project manager and the performing organization management. The project sponsor, a member of the management team, is the highest escalation point and the ultimate decision-maker for the project for the project manager through project planning, execution, and closeout.

The project sponsor and the project manager must meet at the start of project initiation to discuss and share a single vision for the project. They both need to establish a solid bilateral communication, which is later extended to core team members, as well as to key stakeholders involved in the project throughout its development life cycle.

In some organizations, certain types of projects, especially internal service projects and new product development projects, are initiated informally, and some limited amount of work is done to secure the approvals needed for formal initiation. Projects are then typically formally authorized.

2.4.2 Decision on Project Launch: The Project Charter

The project charter is a formal document that authorizes the project manager to utilize corporate funds and resources to fulfill the project objectives, or in some cases to only reach the project initiation stage gate. The charter demonstrates management support for the project, authorizes the project manager to lead the project and allocate resources as required. The project charter is signed and sanctioned by the project sponsor. It states the name and purpose of the project, the project manager's name and roles and responsibilities, the identified key stakeholders and a statement of support from the sponsor.

The project charter should be distributed to all with an interest in the project—see Section 2.5.1, Stakeholder Management. This will reinforce the project manager's authority.

The performing organization management plays an active role following the issue of the project charter. Decision-makers must be identified so that they can participate in interviews and document reviews. The performing organization management communicates the project history and background to the project manager and the core team, helps define the business origin of the project and its critical success factors and constraints, and commits resources to the project.

The project charter should accurately reflect the vision of the executive management. The project objectives should be explicit as to how the expected outcome of the project will benefit the organization and help it achieve its business goals. Critical success factors (CSFs) should identify the project outcomes that will define the project as a success.

A template for a project charter is shown in Figure 2.9.

2.4.3 Project Initiation Phase Kick-Off Meeting

In conjunction with the project sponsor, the project manager should plan a project initiation phase kick-off meeting, inviting all parties that will be involved in project initiation and the subsequent project phases. The kick-off meeting is the first opportunity for the project manager to meet the key project players, for the project sponsor to discuss the project vision, and is the occasion for the performing organization management to demonstrate support and commitment to the project and advocate project success.

The kick-off meeting should be an instructive and stimulating event, conveying the same vision to all players and ensuring common understanding and alignment. It is also an opportunity for all members of the organization not already participating in the project to become formally involved.

2.4.4 Project Governance after Launch

The project manager should identify any governing body or steering committee to which the project is accountable, and how the project manager is accountable to these and other key groups.

Project Charter - Template

Project Charter	
Project Name	
To	Distribution List
From	Initiating Authority
Date	
Assignment	(*Project Manager Name*) is authorized as project manager for the (*Project Name*) effort. (*Project Manager Name*) is designated to ensure customer satisfaction and to manage the project to a successful conclusion, as described in this charter. (*Project Manager Name*) will be responsible for internal communication and cooperation with functional managers included in the distribution list.
Responsibility	(*Project Manager Name*) will: • Be the primary point of contact for (*Internal Organization*) and (*Customer Name*) • Prepare a detailed project plan, and get agreement to that plan from the corresponding functional managers • Ensure team members know their responsibilities • Track team member performance • Track overall project performance • Report project status to management • Maintain a project binder containing all pertinent project data • .....
Authority	(*Project Manager Name*) 's authority includes: • Authority to direct the project team • Access (*Customer Name*) to on all matters related to this effort • Access to the functional managers on all matters related to this effort • Control of the project budget • Access to financial reports related to project expenditures, including time and attendance • Renegotiations with functional managers to delegate responsibility and authority to functional organization team member • .....
Scope Statement	

Signature	Name	Function

FIGURE 2.9
Project charter template

2.4.4.1 The Project Steering Committee

The project steering committee is comprised of members of the performing organization management team, chaired by the sponsor, who meet on a regular basis to evaluate the project through the development life cycle

2.4.4.2 *Decision-Makers*

Decision-makers are members of the organization who have been designated to make project decisions on behalf of major business units that will use, or will be affected by, the product or service the project will deliver. Decision-makers are responsible for obtaining consensus from their business unit on project issues and outputs, and communicating it to the project manager. They attend project meetings as requested by the project manager, review and approve process deliverables, and provide subject matter expertise to the project.

2.4.4.3 *Key Internal Stakeholders*

Individuals or business units, within the performing organization, who are the most affected by the new product or service are termed key internal stakeholders.

2.5 STAKEHOLDERS

2.5.1 Stakeholder Management

Stakeholder management is a key skill for all project managers. Stakeholders are individuals who represent specific interest groups served by the outcomes and performance of a project or program (see Figure 2.10). Project managers are accountable for the end-to-end management of their projects, including performance and expectation management of individuals who may be outside their direct control.

"Individuals and organizations who are directly or indirectly involved in the project, or whose interests may be positively or negatively affected as a result of project execution or successful project completion."

FIGURE 2.10
Stakeholders—a definition

Project managers must give due consideration to the people issues surrounding projects and recognize that the appropriate involvement and management of stakeholders is a critical success factor. Project managers should therefore institute a formal stakeholder management process that is appropriate for the nature of the project.

The degree to which a stakeholder is required to contribute to a project to ensure success depends on a number of factors, including the stakeholder's position and authority within the organization, the degree to which the project relies upon the particular stakeholder to provide a product or service, the level of "social" influence of the individual and the degree to which the individual is familiar with specific aspects of the business. Each of these factors can be analyzed in more detail during stakeholder analysis.

The project manager should make an explicit effort to understand the full extent of the project, and cast wide to capture all potential stakeholders who have some interest or level of influence that can impact the project. Stakeholders are not confined to the internal performing organization. The PESTLE model in Figure 2.11 is widely used as a framework to scope stakeholder involvement.

When establishing the stakeholder relationship, the project manager must be aware of and understand that stakeholders are human and have

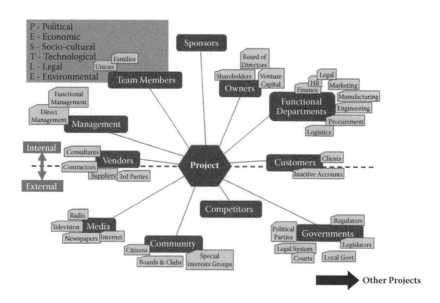

FIGURE 2.11
The PESTLE model

FIGURE 2.12
Identifying stakeholders

personal goals and political motivations. Their professional and personal interests will influence decision making. Their key motivations, however, are professional, and their concerns will differ at different stages of the project. In some cases, for some stakeholders, their interests will often go beyond the visible project's goals.

For corporate projects that are extensive across many levels of the organization, and are spread across many geographies, the project manager will deal with a large number of internal stakeholders and must consider the multiple tiers of stakeholder involvement. Using the model illustrated in Figure 2.12, the project manager can structure and categorize the stakeholders in expanding concentric circles.

Primary stakeholders are immediate communities of interest and can be marked inside the first circle. Secondary stakeholders are the intermediaries in the process, and so on to the outer rim of the circles. The project manager in this way will highlight many groups that do not think of themselves as stakeholders.

Stakeholders can also be categorized by their influence across the total time frame of the ownership/operational cycle. Key stakeholders who are behind the incentive of the project and who have the highest expectations as to the strategic and/or operational benefits and business value are in the "Drivers" category. Stakeholders who perform on the project development are the "Doers." Stakeholders who are responsible for the

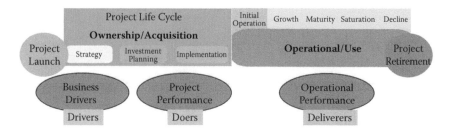

FIGURE 2.13
3D stakeholders—along the life cycle

operational use of the resulting product or service are the "Deliverers" of the business value. This "3D" approach is illustrated in Figure 2.13.

2.5.2 Stakeholder Analysis

Stakeholder analysis is performed to identify and build a comprehensive list of the project's stakeholders, to assess their interests and to determine how they affect the project viability and performance. The analysis establishes the goals and roles of different groups and formulates appropriate forms of engagement with these groups.

A stakeholder analysis is the first step in building the relationships needed for the success of a project. Importantly, it establishes the social environment in which stakeholders operate and defines the approach to achieve this. A stakeholder analysis will:

- Identify and define the characteristics of stakeholders
- Determine the interests of stakeholders in relation to the problems that the project is seeking to address
- Discover conflicts of interests between stakeholders
- Identify relations between stakeholders that may enable "coalitions" of cooperation
- Assess the capacity of different stakeholders to participate
- Establish the appropriate type of participation by different stakeholders

Stakeholder analysis is a five-step iterative process (see Figure 2.14):

1. Identify project stakeholders
2. Identify stakeholders' interests

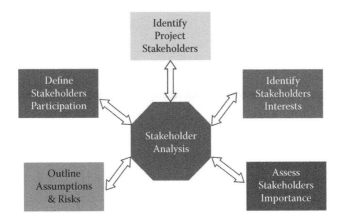

FIGURE 2.14
Stakeholder analysis synoptic

3. Assess stakeholders for importance and influence
4. Outline assumptions and risks
5. Define stakeholder participation

2.5.2.1 Identify Project Stakeholders

Identification of the project's stakeholders is a process of developing a list of those who are impacted by the project or will impact the project, positively or negatively. Core project team members and individuals from the organization familiar with the project's deliverables and constraints, and the organization's structure and politics, should be involved in developing the list. Identification is best done using the PESTLE model. Stakeholders can be individuals, groups, communities, organizations, etc. Breaking stakeholder groups into smaller units will often assist in identifying important groups who may otherwise be overlooked. The list can be expanded following a brainstorming session or any other process commonly used in the organization.

The process of identification of stakeholders involves the participation of individuals who may have divergent views about the identified stakeholders; often there is a need to challenge views and agreements made during this process.

After the stakeholders have been identified, these can be categorized referring to the PESTLE model illustrated earlier in Figure 2.11. Another useful graphical categorization is the Stakeholder Wheel shown in Figure 2.15.

FIGURE 2.15
The stakeholder wheel

PRACTICE EXERCISE

Develop the stakeholder list for your current project, or a completed project.

2.5.2.2 Identify Stakeholders' Interests

The project manager schedules and holds meetings and interviews with the stakeholders listed during the first step of analysis. Discussions should be open and frank to draw out the key interests for each stakeholder in relation to the problems that the project is seeking to address.

For best results, the process of identification should be conducted as a series of facilitated workshops. Through this, knowledge about stakeholders, their interests, power, and influence can be uncovered and documented.

Individually the team members will benefit from exposure to new ways of understanding relationship management, and they will learn about the characteristics, leadership and management styles, and expectations of the project's key stakeholders.

From the stakeholder categories, or other grouping techniques, the project manager can build the context of the meetings by structuring questions that concern the target category. For example, business managers are more concerned about financial and organizational issues, whereas technology experts will focus more on infrastructure and processing issues. The PESTLE interest spectrum in Figure 2.16 can be used as a guideline to structured questions.

The project manager should conduct discussions by inviting the stakeholders to express themselves regarding how they perceive the project and what guidance they could offer.

Key questions could include:

- What are the likely expectations of the project by the stakeholder?
- What role would the stakeholder wish to hold during the project?
- What benefits are there likely to be for the stakeholder?
- What negative impacts may there be for the stakeholder?

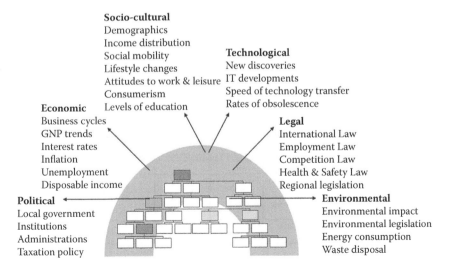

FIGURE 2.16
PESTLE organizational impacts

Stakeholder						Scale		
Organization	Title	Quality	Name	3D Type	Interests	Importance	Influence	Priority
Internal	VP xxx	Sponsor	J. Top	Driver				
Internal	Sr. PM	Project Mgr	K. Nimble	Doer				
Internal	Prod. Mgr	Ops Mgr	A. Giver	Deliverer				
Supplier X	Sales Mgr	Project Mgr	E. Deal	Doer				
.....								

FIGURE 2.17
Stakeholder interests

- What resources is the stakeholder likely to commit (or avoid committing) to the project?
- What other interests does the stakeholder have that may conflict with the project?
- How does the stakeholder regard others on the list?

The stakeholders' interests are recorded (Figure 2.17).

2.5.2.3 Assess Stakeholders for Importance and Influence

This step assesses the importance and influence of each stakeholder on the project. A scale from 1 to 5 can be used for each assessment, where 5 is the most important or most influential.

Importance is scaled to represent how much the stakeholder's issues, expectations, needs, and interests are related to the aims of the project. If the highest-scaled important stakeholders are not involved or assisted, then the project will be on track for failure.

Influence is scaled to represent both formal and informal power held by the stakeholder (see Figure 2.18). Hierarchical positions, such as project

Stakeholder						Scale		
Organization	Title	Quality	Name	3D Type	Interests	Importance	Influence	Priority
Internal	VP xxx	Sponsor	J. Top	Driver				
Internal	Sr. PM	Project Mgr	K. Nimble	Doer				
Internal	Prod. Mgr	Ops Mgr	A. Giver	Deliverer				
Supplier X	Sales Mgr	Project Mgr	E. Deal	Doer				
.....								

FIGURE 2.18
Stakeholder importance

sponsor, management or project steering committees, operational managers, and primary users would score high. This would also be the same for informal influencers, such as a project champion, a subject matter expert, or key external provider.

The scores are reviewed, agreed on, and recorded by the project team. A first-order priority and ranking is then made by calculating the product of the importance and influence scores.

PRACTICE EXERCISE

From your previously developed stakeholder list, complete the table in Figure 2.19. (If there are not enough lines in the table, please use the space below.)

Stakeholder						Scale		
Organization	Title	Quality	Name	3D Type	Interests	Importance	Influence	Priority

FIGURE 2.19
Stakeholder interests template

2.5.2.4 Outline Assumptions and Risks

The exercise of analyzing and mapping the stakeholder base will improve the project manager's insight into the stakeholders and their drivers. The importance/influence mapping will indicate how each individual stakeholder relates to the project's goals and requirements.

The project manager and the project team can now focus on the assumptions relating to both the project initial scope and the stakeholder's expected consistency throughout the development life cycle. On medium- or long-duration projects, and/or complex internal/external interrelationships, changes in the composition of important stakeholders and their expectations are to be anticipated and the underlying assumptions recorded.

The risks assessed in the stakeholder analysis refer more to the nature of the stakeholders, the level of their interests, the viability of their expectations, and how they rank in priority. All these key factors will place uncertainty on the stability of the project scope and thus impact the project's success. For example, low-interest stakeholders who hold high importance and influence can resist the project's goals and create obstacles outside of the control of the project manager.

2.5.2.5 Define Stakeholder Participation

This step closes the cycle, with the project manager and the project team completing an assessment of the capacity of different stakeholders to participate and determining their appropriate type of involvement at successive stages of the project cycle.

The project manager should plan strategies for approaching and involving each individual or group of stakeholders. Special attention should be given to reluctant stakeholders, while a specific monitoring mechanism should be instituted for those stakeholders who may change their level of involvement as the project is performed. When the stakeholder is a group rather than an individual, the project manager will need to decide whether all members of the group participate or only selected representatives of the group.

The project manager can establish a first scheme of involvement using the importance/interest grid in Figure 2.20.

- High Importance and High Interest: Fully *engage* these stakeholders and make the greatest efforts to satisfy them
- High Importance and Low Interest: Place enough effort to *satisfy* these stakeholders, with no excess to avoid weariness of message

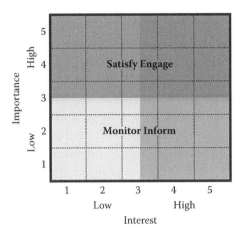

FIGURE 2.20
Stakeholder level of interest

- Low Importance and High Interest: *Inform* these stakeholders adequately, to ensure that no major issues arise. Utilize these stakeholders sensibly as often they provide assistance and help with the detail of the project
- Low Importance and Low Interest: To *monitor*, with no excessive communication.

The matrix maps the stakeholders according to their ability to influence the project's success or failure. The project manager must seek to appreciate and recognize the stakeholder interests and also establish an involvement status according to their "interest duration" (i.e., continuous, frequent, intermittent or sporadic, or once-off).

The project manager and the core team should then complete the stakeholder analysis by defining for each stakeholder their appropriate involvement, their level of support, and their receptiveness to messages about the project. The team can then proceed to draft the stakeholder communication plan.

2.5.3 Stakeholder Communication Plan

Frequent and comprehensive communication is a key project success factor. The stakeholder analysis provides the expectations of the stakeholder and identifies what each stakeholder requires from the project. The stakeholder communication plan must illustrate how the information needs of all project team members and other stakeholders will be satisfied and verified with a feedback loop. The plan defines for each stakeholder the type of communication most likely to be effective, specifying the timing and frequency, the medium, and the expected outcomes (see Figure 2.21).

Knowing the category of a stakeholder will provide the key to how the communication plan should be developed and delivered. For example, communication to a senior manager will need to contain only the information, and in the format necessary, to provide management with essential data about the project, whereas communication to team members will need more detail and a different language.

The project manager will use the stakeholder communication plan to engage in networking with influential stakeholders, to gain and sustain support from advocates, and to achieve buy-in from critics. The project manager must strive to communicate in stakeholders' "language."

The results of the stakeholder communication plan are integrated into the project communication plan and the project schedule. During project implementation, they are monitored through team meetings and regular reports.

PRACTICE EXERCISE

Now complete the stakeholder communication plan using the table in Figure 2.21.

Stakeholder	Objectives	Type of Communication	Medium	Frequency and Schedule	Distribution	Outcome
xxxx	xxxx	xxxxxxxx	xxxx	xxxx	xxxx	xxxx

FIGURE 2.21
Stakeholder communication plan

2.5.4 Pragmatism with Stakeholders

The project manager cannot expect to align all stakeholders all the time. The project's stakeholder community changes as stakeholders move within the organization or leave it. Consequently, new stakeholders have to be included, and current stakeholders will experience changes to their relative importance to the project, as well as their power and influence. As the project proceeds through its life-cycle phases, different stakeholders may have more or less of an impact, and as a corollary, their levels of importance, and communication requirements will change.

The key issues to address are:

- Is the stakeholder communication plan as effective as it should be?
- What additional actions should be taken to gain more support from the project advocates?
- What are the strategies to implement to convince critics?
- Does the stakeholder importance/interest matrix reflect the current state?

The project manager will also need to establish a monitoring scheme to capture:

- Who influences the stakeholders' opinions generally?
- Who else might be influenced by their opinions?
- Do influencers therefore become important stakeholders?
- If the influencers are not likely to be positive, what will be required for them to support the project?
- If there is little possibility or probability of gaining their support, how will their opposition be managed?
- What is their current opinion of the project manager?
- Who influences their opinion of the project manager?

Because of the dynamics of the stakeholder community, the effectiveness of the stakeholder analysis and the stakeholder communication plan will suffer if not adapted to the evolving situation. The project manager must therefore plan for and prepare the project team to repeat the stakeholder management process in its totality, or in part, many times as the project progresses through its life-cycle phases or as the stakeholder community changes.

2.6 PROJECT REQUIREMENTS

2.6.1 Scope Management

The project scope corresponds to the business needs; it links specific deliverables to the business benefits and fulfills stakeholders' expectations.

Scope management consists of the processes required to ensure that the project includes all the work required, and only the work required to complete the project successfully. Project scope management is primarily concerned with defining and controlling what is included in the project and what is excluded. The process includes the following.

2.6.1.1 Specification of Scope

- Defining and documenting stakeholders' needs to meet project objectives
- Developing a detailed description of the project and product/service
- Subdividing project deliverables and work into smaller more manageable components

2.6.1.2 Management of Scope Changes

- Monitoring the status of the project and product scope and managing changes to the scope

2.6.1.3 Verification of Scope

- Formalizing acceptance of the completed project deliverables

The word "scope" used without any qualification is often confusing as it fails to describe the contents and the state of the project context. For example, at start of project, the scope is conceptual and generalized. As the project nears commissioning and handover, the scope is detailed and materialized.

The word "scope" can be qualified as follows to describe its level of detail and evolution:

- **Scope statement**—describes the major objectives and states the project deliverables
- **Scope description**—describes the needs, main theme, and key components of the project

- **Scope definition**—breakdown of the project corresponding to the functional requirements
- **Scope specification**—comprehensive design of the project to be delivered
- **Scope of work**—detailed definition of work to be performed under a contract or subcontracted for the completion of the project. Also called Statement of Work – SOW

2.6.2 Specification of Scope

The *project scope* refers to the work to be accomplished to deliver a product, a service, or a result with the specified features and functions. The *product scope* defines the features and functions that characterize a product, a service, or a result. Scope planning is the process of progressively elaborating and documenting the project work that produces the product of the project. This progressive elaboration will not only break down the problem to as low a level of detail as possible, it will also address the inherent unknowns identified at the start of the project.

As was illustrated in Chapter 1, the Engineering V project development life cycle presents the backdrop to scope planning (see Figure 2.22).

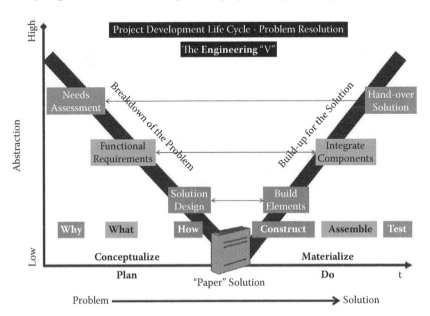

FIGURE 2.22
The engineering V

The conceptualization portion of the Engineering V covers the initiation and planning phases and corresponds to:

- Defining and documenting stakeholders' needs to meet project objectives
- Developing a detailed description of the project and product/service
- Subdividing project deliverables and work into smaller more manageable components

The project objectives drive the context of the project. Objectives are characterized by the five elements of the SMART model:

S = Specific
M = Measurable
A = Agreed-upon
R = Realistic
T = Time-constrained

Drafting project objectives is an exercise in linguistic and grammatical skills. Each objective is a phrase with a meaningful verb, a noun, and a qualifying statement. The verb states a completed action, the noun the actual result, and the qualifying statement will describe any or all of the three vectors of the triple constraint: scope, time, and cost. *Place a new A/C unit in the office* is not a SMART objective. The SMART objective would be *Install a functionally approved air conditioning unit according to the technical specification ABC/123 by the 15th June yyyy within the agreed cost of $15,000.*

2.6.3 Key Documents

During the conceptualization step, three key documents are produced: needs assessment, functional requirements, and solution design.

2.6.3.1 Needs Assessment

The needs assessment document is a record of the stakeholder needs that motivate the development of the project solution. It is essential that these needs be well understood and agreed upon before the initial phase begins. Needs are often vague, ill formulated, or not stated. The needs assessment process clarifies these needs and is a record of the stakeholders' expectations in a clear and complete manner.

Generally, it is not possible to meet all of the needs within the time and budget available for the project. Often the stakeholders may have conflicting needs. Tradeoffs and prioritizations need to be made to balance the needs that will be the focus of the project.

There are several purposes for the needs assessment:

- Get and document stakeholder agreement on the needs that the project is to meet to ensure that the development starts off in the right direction, to avoid later redirection
- Clearly describe the needs that the project will meet, as the first step toward defining scope requirements
- Document the process and results of stakeholder consensus, relative to conflicting needs
- Demonstrate to the stakeholders that their individual views have been incorporated

The needs assessment template can be adapted to the size and nature of the project.

NEEDS ASSESSMENT TEMPLATE

Section	Contents
1.0 Purpose of Document	This section is a brief statement of the purpose of this document. It is a description and rationale of the needs that the project will meet. This is a vehicle for stakeholder feedback, and a justification for the key needs selected
2.0 Overview	This section gives a brief overview of the project, describes the stakeholders, and the expected role of each (expanded in the Stakeholder Analysis)
3.0 Referenced Documents	This section lists supporting documentation used and other resources useful in understanding the performing organization.
4.0 Needs Collection	This section describes the needs collection process used. Records of interviews and their results are included.
5.0 Needs Description	This section is the heart of the document. It describes clearly and fully the needs expressed by the stakeholders. The essential needs are highlighted, and distinguished from those that are "wishes" that may be sacrificed for cost or for more critical needs. The needs have corresponding measures of effectiveness or measures of performance that provide metrics for achievement.
6.0 Needs Validation	This section describes the process and results of validating the collected needs with the stakeholders.

7.0 Gap Analysis	This section describes the needs and compares it to the current situation. Gaps are identified and ranked by criticality and importance.
8.0 Needs Comparison	This section describes conflicting needs. Options are compared on estimated life cycle cost, ease of implementation, or other agreed-upon criteria.
9.0 Prioritization of Needs	This section defines the process and results of prioritizing the needs, and the rationale for the selection.
10.0 Validation of Key Needs	This section documents the stakeholders' agreement that the project will focus on the prioritized needs.
11.0 Appendix	The appendix includes results and other documents which substantiate the needs assessment.

2.6.3.2 Functional Requirements

The functional requirements document elaborates on the needs assessment and describes what the product/service (system) will perform, how well it is to perform in the operational environment, and under what conditions. This document does not define how the system is to be built and is written in a language style understandable to stakeholders. This document sets the technical scope of the product/service to be built. It is the basis for verifying the system and subsystems when delivered. The functional requirements document defines all the major system functions, and describes for each a set of requirements: what the function does, and under what conditions, e.g., environmental, reliability, and availability. Each requirement is linked and traceable to a stakeholder-specified need and is written in a concise, verifiable, clear, feasible, necessary, and unambiguous manner. Each requirement will stipulate the method that will be used to verify it.

The functional requirements template can be adapted to the size and nature of the project.

FUNCTIONAL REQUIREMENTS TEMPLATE
Adapted from the IEEE Std 1233

Section	Contents
1.0 Scope of System or Subsystem	Contains a full identification of the system
	Provides a system overview and briefly states the purpose of the system
	Describes the general nature of the system
	Identifies the project stakeholders
	Identifies current and planned operating environments

2.0 Reference	Identifies all needed standards, policies, laws, concept of operations, concept exploration documents, and other reference material that supports the requirements
3.0 Requirements	Functional requirements—What the system will do
	Performance requirements—How well the requirements have to perform
	Interface requirements
	Data requirements
	Nonfunctional requirements, such as reliability, safety, environmental
	Enabling requirements—production, development, testing, training, support, deployment, and disposal
	Constraints—technology, design, tools, and/or standards
4.0 Verification Methods	For each requirement, identify one of the following methods of verification:
	Analysis—by reviews or mathematical analysis
	Test—by comprehensive testing
	Inspection—verification through a visual comparison
5.0 Supporting Documentation	Sources of information that add to the understanding of the requirements: diagrams, analysis, memos, rationale
6.0 Traceability Matrix	This is a table that traces the requirements to the user requirements or needs
7.0 Glossary	Terms, acronyms, definitions

2.6.3.3 Solution Design

The solution design document defines the architecture, components, modules, interfaces, and data for a system to satisfy the specified functional requirements. The document will take many different forms and vary between organizations. The solution design is the last major product document and triggers the completion of the project plan. (See Chapter 3.)

The project requirements and design documents progressively elaborate on the needs assessment to produce:

- The characteristics of the product/service
- The project deliverables
- The project assumptions
- The project exclusions

- The project constraints
- The project acceptance criteria

During the development of functional requirements and solution design, the project manager can utilize prototyping, proof of concept, and/or progressive elaboration to help stakeholders understand and visualize how needs may be fulfilled. This is discussed in detail in Chapter 3.

2.6.3.4 Impact on the Project Scope

As the needs are broken down and as the requirements and design documents are elaborated, the project scope will increase in detail and volume to be finalized as the scope specification or scope of work. Stage gates and approvals are placed at milestones between the needs and requirements, the requirements and design, and once the design is completed. This allows for alignment to the original project intents and incorporates any required internal or external scope adaptations. The stage gates are also opportunities to review the previously allocated priority to the project, and may lead to a suspension or premature termination of the project.

2.6.3.5 Assumptions and Constraints

The project assumptions are established, validated, and documented from the set of identified unknowns that cannot be discovered during the conceptualization step. The project assumptions as they stand at the end of this step reflect the level of detail attained for the scope specification, the nature of the product scope, and the acceptable comfort level of the performing organization management. That comfort level will lead to the decision to proceed with the project, to extend the conceptualization step to resolve key assumptions or to terminate the project because of the amount of unknowns that can jeopardize the project success.

2.6.4 Management of Scope Changes

Scope change management is concerned with influencing the factors that create scope changes to ensure that changes are agreed upon, determining that a scope change has occurred, and managing the actual changes when and if they occur. Scope change management must be thoroughly integrated with the other project control processes.

Management of scope changes is discussed in detail in Chapter 7.

2.6.5 Verification of Scope

Scope verification is the process of obtaining formal acceptance of the project scope by the stakeholders. It requires reviewing deliverables and work results to ensure that all were completed correctly and satisfactorily. If the project is terminated early, the scope verification process should establish and document the level and extent of completion. Scope verification is primarily concerned with acceptance of the work results.

Verification of scope is discussed in detail in Chapter 8.

PRACTICE EXERCISE

Assess your current project documentation. Write the SMART objectives and give a short description of the needs, functional requirements, and solution design.

2.7 ROLE OF THE PROJECT MANAGER

2.7.1 Project Manager Formal Assignment

The project manager is formally assigned at the start of the project initiation phase. The project sponsor approves the project charter in which the role and responsibilities of the project manager are established, and authority to utilize corporate funding and resources is given.

The project manager will facilitate and coordinate the project team from the initiation phase to close-out and handover, and oversee the team's performance of project tasks. It is also the responsibility of the project manager to secure acceptance and approval of deliverables from the project sponsor and stakeholders.

2.7.2 Organizational Structures

One of the first tasks for the project manager is to break the project into its major components (see Figure 2.23). This first structure is built using a work breakdown structure (WBS), which represents how the project should best be organized and how subprojects are composed. (The WBS tool is detailed in Chapter 3.) It is sufficient at this stage to know that the high-level WBS corresponds to the organizational chart of the project and identifies the constituent parts of the project core team. This first draft of the organizational chart depicts the functional structure of the project and will usually be projectized (as already discussed in Chapter 1).

The projectized project organization is ideally suited to allow the project manager full management and control of assigned resources. However, the performing organization's culture, style, and structure will force the project manager to adapt this first-cut chart. Furthermore, the company's functional structure will influence the project and impose a matrix or cross-functional organization. The aforementioned factors, plus other factors concerning the integration of external providers and/or partners and joint-venture, will impose on the project manager an organization that distances itself from the original desired structure.

The project manager must strive to establish an organizational structure that provides for the management and control of resources. This entails allocating key resources to perform solely for the project, without interruptions from normal operational responsibilities or other projects.

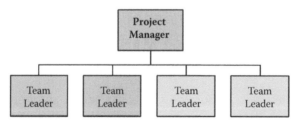

FIGURE 2.23
Basic project structure

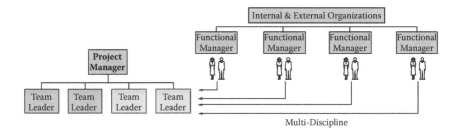

FIGURE 2.24
Cross-functional assignments

However, in the majority of cases, the project manager will have to settle for a cross-functional matrix organization, in which resources are allocated from different parts of the performing organization (see Figure 2.24). The individuals assigned will then have double-reporting lines to both the functional manager and the project manager.

The project manager will have to consider the impact this creates on the assignment and availability of resources, and establish bilateral agreements with the providing functional managers. When the assigned resources are physically located with the providing functional manager, the project manager will have little or no influence on the allocation of work activities as the project schedule will be driven by the functional manager's power of decision. This is compounded when the resources are geographically dispersed. This situation produces a weak matrix structure for the project manager and places great difficulties on achieving the project goals.

2.7.3 Project Core Team Allocation

Whatever organizational structure is chosen or imposed, the project manager must establish a project core-team as soon as possible. Individuals assigned to this core-team will be delegated responsibility for the key components defined in the high-level organizational chart. The project manager will guide, facilitate, and coordinate the work performance of the assigned team leaders.

Project core-team members will be responsible for breaking down their respective components and developing their subproject plans to produce the corresponding deliverables. As team leaders, they will also provide technical and task leadership for the resources assigned at a later stage.

2.7.4 Engaging the Organization

Throughout the initiation phase, the project manager will focus on establishing key relationships and engaging decision-makers and influencers.

The project manager must go beyond just obtaining approval of the project charter and establish an open and bilateral communication channel with the project sponsor who is the ultimate decision-maker for the project. Both sponsor and project manager must agree on the most appropriate scheme to secure spending authority, to approve major deliverables, and to sign off on approvals to proceed to each succeeding project phase. The sponsor must also agree to provide support for the project manager and assistance in the allocation of resources for the project when the project manager encounters difficulties.

The project manager must establish the means to capture and maintain the expectations of key stakeholders who represent the business units that identified the need for the product or service project. The relationship extends to the decision-makers who have been designated to make project decisions on behalf of major business units that will use, or will be affected by, the product or service the project will deliver. Decision-makers are subsequently responsible for obtaining consensus from their business unit on project issues and outputs, and communicating it to the project manager. The project manager will provide a schedule of project meetings for the review and approval of deliverables.

The project manager must seek engagement and commitment from functional managers of the performing organization. These managers are the primary providers of resources who will perform on the project to produce assigned deliverables.

The operational organization must be included in the project manager's radar, as it will use the product or service that the project is developing. Because of the new product or service, the operational organization will be affected by the change in work practices, by modified workflows or logistics, by the quantity or quality of newly available information, and other similarly operational impacts.

3

Project Planning

3.1 CHAPTER OVERVIEW

This chapter covers the planning phase of a project. The planning and scheduling steps are described, and the specific scope development actions are explained.

This chapter concentrates on the major planning techniques and tools, namely work breakdown structure and estimation, paving the way for the scheduling techniques that are described in Chapter 4.

A case study is introduced as well, along with a set of exercises that will guide the reader through all the steps of project planning.

3.2 SYNOPSIS AND APPROACH

The purpose of project planning is to define the detailed parameters of a project and ensure that all the pre-requisites for project execution and control are in place. The major deliverable of project planning and scheduling is a project plan presented to management for subsequent approval.

Project planning builds upon the work performed during project initiation. The project definition and scope are validated with appropriate stakeholders, starting with the project sponsor and customer decision-makers. Project scope, schedule, and budget are developed, refined, and confirmed, and risk assessment activities advance to the mitigation stage. The initiation deliverables and initial project plan are further developed, enhanced, and refined, until they form a definitive plan for the rest of the project.

Core project team members are brought on board as early as possible in the planning process, and familiarized with the project objectives and environment. Additional resources may be requested for the finer development details of the project plan.

Project planning is an opportunity to identify and resolve any remaining issues and answer outstanding questions that may undermine the goals of the project or threaten its success. It is an opportunity to plan and prepare, as opposed to react and catch up.

Project sponsorship and commitment are reconfirmed at the end of the phase, with approval signifying authorization to proceed and commit funds for project execution and control.

3.2.1 Challenges to Project Planning

The end result of the project planning stage, and the subsequent project scheduling, is a project plan that management can validate and approve. They can subsequently release funds and allocate resources for the project execution phase. Following management sign-off of the project plan, a project baseline will be established and the "organization wheel" is set in motion, triggering actions in the various departments and functions of the company to make the organization ready to receive the project's outcome.

For commercial projects, the approved project plan is built into the formal proposal and bid to the client/customer.

Thus the approved project plan becomes a major commitment and the project manager must fully appreciate that its contents must be feasible, viable, realistic, and achievable.

The project manager is, however, faced with a series of challenges during the planning stage.

The major challenge is reaching a detailed scope specification and a scope of work. This entails ensuring that the product design is sufficiently detailed to allow for detailed project execution planning. This is discussed at length in Section 3.5, Scope Development.

The Engineering V has been previously introduced and acts as a backdrop to the detailed breakdown.

Another major challenge is raising the level of knowledge of both problem/need and proposed paper solution (see Figure 3.1). As the scope is broken down, the level of the knowledge base increases as the project team analyzes the problem to reach a satisfactory solution. This reduces progressively the "unknowns" (which may increase as more knowledge is

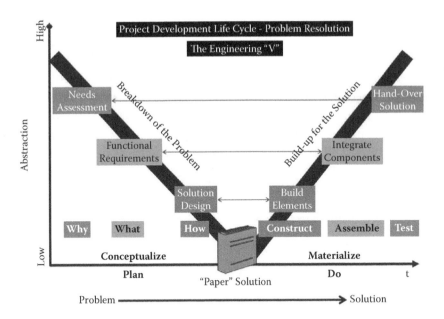

FIGURE 3.1
Project development life cycle—problem resolution

acquired). As illustrated in Figure 3.2, at whatever point in time project planning is judged to be complete, there will always be unknowns.

The aim of the project manager is to balance judiciously the level of detail to reach and the time and resources required to reach that level.

The project manager must guide the project team to establish thorough documentation that reflects known "real" facts of the current state of the scope. Related to this challenge, and equally important, is determining

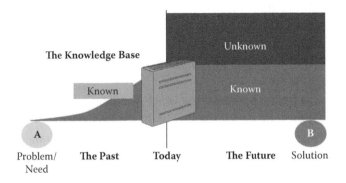

FIGURE 3.2
The knowledge base

and understanding the volume of assumptions that are made from the identified unknowns, and validating their characteristics. Assumptions must be clearly described as they are considered to be "virtual" facts in the elaboration of the project plan. Both "real" and "virtual" facts make up the committed scope of work.

As is explained fully later in this chapter (see sections 3.6 and 3.7), the project manager and the team will convert the scope into work to be performed and estimate the duration and cost of each activity so as to construct the overall project schedule and total project cost. The challenge for the project manager is to recognize the range of accuracy of all estimates and to make this clear to management at project plan approval.

Projects are performed in an environment of uncertainties. No project exists without risks (discussed in Chapter 5, Risk Management). As another major challenge, the project manager and the team must incorporate the results of mitigation and contingency plans in the project plan.

Last and not least, the project manager and the team must clearly identify the nature and amount of resources that need to be allocated to the project during project execution. The project manager must secure commitments for resource assignment and release, according to the project plan, from internal functional managers, external suppliers, and eventually the client organization.

3.2.2 The Project Planning Environment

At a very high level of generalization, project planning is the process of converting the project scope into work to be performed and to quantify that work so as to arrive at a schedule and a budget.

Project planning is a systematic and structured approach to create a viable and realistic project plan and build a commitment for subsequent project execution.

Figure 3.3 presents the basis from which the detailed project plan can be elaborated. It must be noted that this is an iterative process, which follows the breakdown of the problem into a workable paper solution.

The current content of the project scope will drive the organizational structure of the project. The project manager should seek to utilize company or industry-recognized standards and methodologies to assist the project team to benefit from previous experiences and establish the most

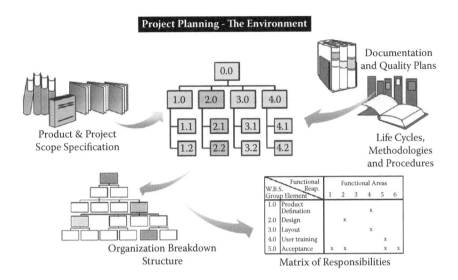

FIGURE 3.3
Project planning—the environment

appropriate organization and division of work. (Further explained in Section 3.6, Work Breakdown Structure.)

The project manager must also seek to engage the organization, through the sponsor and stakeholders, to release and assign pertinent resources to the project planning phase. From the negotiations to secure the resources, a matrix of responsibilities can be established.

3.3 ESTABLISHING THE PROJECT CORE TEAM

The project manager's primary step, before constituting the project core team (Figure 3.4), is to partition the project in segments that will become the subprojects. This should have been done during the project initiation phase at the approval gate to move into the planning stage.

The partitioning will illustrate how best to organize the project: by major deliverable, by discipline, by functional department or any other suitable division (see Figure 3.5). (For further explanation, see Section 3.6, Work Breakdown Structure). The aim is to set up the initial project organization structure as early as possible, and staff it from the internal organization or, when necessary, from external organizations in the case of contracting or outsourcing.

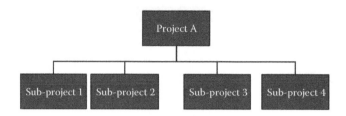

FIGURE 3.4
The project core team

The project manager must ensure that core team members are released by their management and assigned to work together through the major steps of the project planning. Line managers who have given their approval for resource availability, at end of project initiation, must fulfill their commitment; if not, the project manager will be unable to proceed with the project planning, as the project will lack expertise.

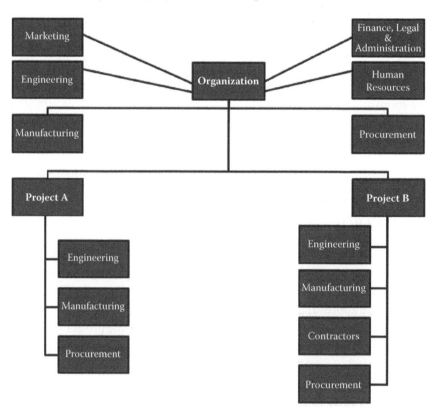

FIGURE 3.5
Project structure by discipline/functional group

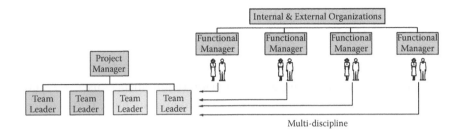

FIGURE 3.6
Team lead assignment

As illustrated in Figure 3.6, core team members are to have the competence and experience in the major segments of the project that will be assigned to them to lead.

The project manager must seek to enhance the ability of core team members to contribute quickly and positively to the project's desired outcome (see Figure 3.7). A clear set of responsibilities and accountability must be agreed on with each core team member, so that coherence is maintained in the newly constituted project team.

A project planning kick-off formally marks the beginning of project planning and facilitates the transition from project initiation. It ensures that the project remains on track and focused on the original business need. The project manager will brief the core project team members thoroughly, distribute the on-hand project documentation to date, review the current project status, and reexamine all prior deliverables produced during project initiation.

Information to be made available to new team members includes:

- All relevant project information from project origination and initiation
- Organization charts for the project team and performing organization

FIGURE 3.7
Core team project organization

- Information on project roles and responsibilities
- Project procedures (team member expectations, how and when to report project time and status, sick time and vacation policy)
- General information about the customer logistics (parking policy, work hours, building/office security requirements, user ID and password, dress code, location of restrooms, supplies, photocopier, printer, fax, refreshments, etc.)

The project manager (or team leader, if appropriate) must convey to each new team member, in a one-on-one conversation, the position's role and responsibilities related to the project. To streamline interaction among the team, new team members must also become familiar with the roles and responsibilities of all other project team members and stakeholders as soon as possible and immediately receive copies of all project materials, including any deliverables produced so far. It is usually the project manager's responsibility to get new members of the team up to speed as quickly as possible. On large projects, however, if the team is structured with team leaders reporting to the project manager, it may be more appropriate to assign a team leader to "mentor" the new individual.

3.4 THE KEY PLANNING STEPS

Project planning is an iterative process performed in parallel with the product problem analysis and solution design (see Figure 3.8). By doing so, both the project and product planning can not only be synchronized throughout the project planning phase, but also can ensure bilateral coherence and self-check. Furthermore, early knowledge of product requirements and design can assist greatly in the project structuring.

The project manager must first of all finalize both the project scope statement and the project objectives. These firmly set the framework of the project planning exercise.

Three main steps are then engaged in an iterative process as more detail is reached. The project is structured to identify and describe the work to be performed—this is the work breakdown structure. Each piece of performed work is quantified to establish its duration and cost—this is the estimation. And the work is then sequenced to establish its logical flow—this is the precedence analysis.

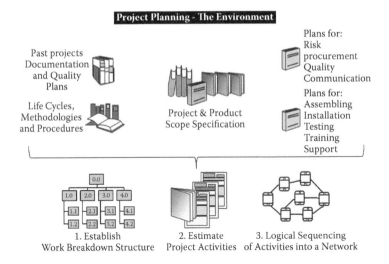

FIGURE 3.8
Project planning—the environment

3.4.1 Finalizing the Project Scope Statement

The initiation phase and the stage exit management approval have established the key elements of the project scope. The project manager should, however, finalize this and expand the scope statement to clearly restate the key stakeholders and decision-makers, and any specific expectations that need to be highlighted.

At this stage of the chapter, a case study is introduced to articulate the major steps to follow during the planning phase.

Workshop 1: Capture the Project's Scope Statement

Today is Wednesday, 6th November.

Read through the Case Study in Section 3.9, then using the form below, complete the major needs and stakeholders of the LEY-WORLD project.

WORKSHOP 1 – Capture the Project's Scope Statement

Project Code & Name	LEY-WORLD – International Trade Fair	
Project Definition Statement		
Project Sponsors		

Stakeholders		
Decision Makers	Maker	Decision
Users		
Specific Expectations		

3.4.2 Finalizing the Project Objectives

Defining project objectives was introduced in Chapter 2.

At the start of the project planning phase, the project manager along with the core team members will review the objectives set during the initiation phase and finalize these and, if necessary, break them down one level for better comprehension.

As a refresher, objectives are characterized by the five elements of the SMART model:

S = Specific
M = Measurable
A = Agreed-upon
R = Realistic
T = Time-constrained

Drafting project objectives is an exercise in linguistic and grammatical skills. Each objective is a phrase with a meaningful verb, a noun, and a qualifying statement. The verb states an action to complete, the noun the actual result, and the qualifying statement will describe any or all of the three vectors of the triple constraint: scope, time, and cost. "Place a new A/C unit in the office" is not a SMART objective. The SMART objective would be "Install a functionally approved air conditioning unit according to the technical specification ABC/123 by the 15th June yyyy within the agreed cost of $15,000".

Workshop 2: Finalizing the Project Objectives

Today is Thursday, 7th November.

Yesterday management accepted the initial scope and charter, and they now expect to sign off on the project criteria at the next management meeting.

The Marketing VP also wishes to assess the profitability of the LZB-Sales venture.

Using the form below, draft the following:

1. The LEY-WORLD project objectives, stated in specific, measurable, and constrained terms
2. Major assumptions and out-of-scope components
3. The project deliverables
4. Key milestone dates
5. Estimated costs—only for LZB-Sales sub-project
6. Estimated revenue—only for LZB-Sales sub-project

WORKSHOP 2 – Finalizing the Project Objectives

Project	LEY-WORLD – International Trade Fair	
Objectives		
Major Assumptions & Out of Scope Components		
Key Deliverables	Key Milestones & Dates	

For PROJECT LZB-Sales ONLY	
Estimated Costs	Estimated Revenue
TOTAL COSTS	TOTAL REVENUE

3.4.3 Work Breakdown Structure

The first step in project planning is to define all the work activities that have to be performed to fulfill the project scope. Many techniques can be used to draft the total set of activities, from a simple "to-do" list to a fully structured hierarchical diagram. The technique that best offers the possibly to "drill" down to the lowest and most appropriate level is the work breakdown structure (WBS). This technique is fully developed in Section 3.6.

The WBS provides a common framework for the natural development of the overall planning and control of a project and is the basis for dividing work into definable increments from which the scope of work can be developed and technical, schedule, and cost performance can be established.

3.4.4 Estimation

The work to be performed as identified in the WBS is estimated to arrive at a duration and a cost for each item. All estimates are approximations. Effective estimating requires a strong understanding of variance in estimating and how to account for/govern this variance.

The project manager must accept and make clear to all team members that estimates are based upon incomplete, imperfect knowledge and assumptions. And most importantly, all estimates have uncertainty. This is fully developed in Section 3.7.

3.4.5 Precedence Analysis

Following work estimation, and prior to proceeding to developing the schedule, the activities are to be sequenced to build the logical flow of work to be performed to fulfill the project objectives.

This is fully developed in Section 3.8.

3.5 SCOPE DEVELOPMENT

A coherently defined project scope is critical to the success of a project. Without a definition that reflects what needs to be performed and the assumptions under which this is performed, work already performed may be subject to rework, resulting in lower team productivity. During project initiation, a basic description of the project and its deliverables was written. Project scope development breaks deliverables into smaller pieces of work, allowing the scope to be more accurately defined. Where the initial project scope statement highlighted the deliverables to be produced in support of the desired project outcome, the developed project scope must go many steps further.

Using the information collected during project initiation, and based upon input gained by communicating regularly with the appropriate stakeholders, the project team must refine the scope statement to clearly define each deliverable—including an exact definition of what will be produced and what will not be produced.

The word "scope" encompasses too many interpretations. When used alone, the word "scope" can refer to whatever state the project is during the life cycle. Qualifying the word will not only provide a better understanding, but will also position in time the evolution of the contents of the total project scope itself.

Figure 3.9 offers qualifiers, using the "scope" of a movie as an analogy.

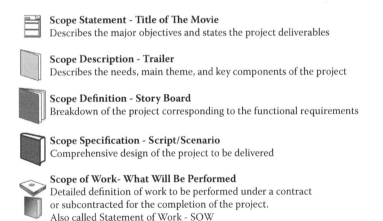

Scope Statement - Title of The Movie
Describes the major objectives and states the project deliverables

Scope Description - Trailer
Describes the needs, main theme, and key components of the project

Scope Definition - Story Board
Breakdown of the project corresponding to the functional requirements

Scope Specification - Script/Scenario
Comprehensive design of the project to be delivered

Scope of Work- What Will Be Performed
Detailed definition of work to be performed under a contract
or subcontracted for the completion of the project.
Also called Statement of Work - SOW

FIGURE 3.9
Scope vocabulary

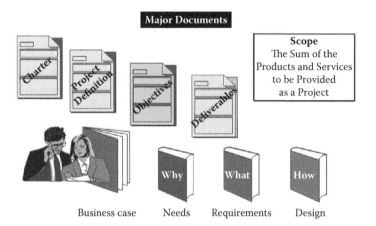

FIGURE 3.10
Scope major documents

With each descending scope qualifier, the volume and detail of the scope contents increase, following the time frame of the project planning phase. Project planning should ideally extend to the scope specification—the script/scenario.

There is a set of key documents that will hold the contents of the scope for both project and product. These are not the only documents that will exist for a project (see Figure 3.10).

During project planning, the project objectives and deliverables set the framework as the project scope definition and specification are developed.

The product scope is developed by proceeding from problem analysis to solution design. This follows the classical engineering steps of: needs analysis—the "Why"; functional requirements—the "What"; and solution design—the "How." These three documents will be expanded, and grow in volume and detail, as more information becomes known about the project.

The project and product scopes are not static—some of the components will continue to change throughout the life of the project. It should be noted that refining the overall project scope occurs in parallel with other project-specific tasks.

3.5.1 Product Scope Development

As was reviewed in the previous modules, projects are driven by the business case (see Figure 3.11). The contents of this document, and principally the objectives to be reached and the deliverables to be produced must be very well understood by all those who will proceed to develop the product scope.

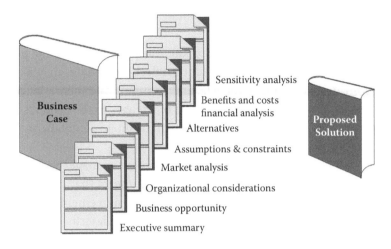

FIGURE 3.11
Business case contents

The nature of the specific line of business associated with the product of each project will drive how the product scope is defined. For example, for building construction projects, architectural drawings will be completed; for application software projects, detailed requirements definition and design will be completed.

As stated previously, product scope development produces three major documents: needs analysis, functional requirements, and solution design (see Figure 3.12). These are reviewed in this section.

Changes to any product scope document must be made using a defined change control process. This process should include a description of the means by which scope will be managed and how changes to scope will be handled. Once documented, the process becomes part of the project procedures. It is vital to document a clear description of how to determine

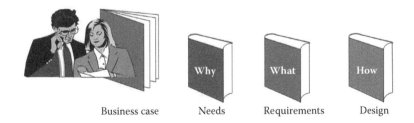

FIGURE 3.12
Major product scope documents

when there is a change in scope to facilitate change control during project execution and control.

Additionally, while updating the product scope documents, the project manager and customer must consider the effect the updates may have on the organization, anticipate impacts, and communicate them proactively to the user community. Selling the value of changes to the product scope during the entire duration of the project will facilitate acceptance at project realization and close-out.

Once again, communication between the project manager and the customer is crucial in creating product scope documents that clearly reflect what the customer needs and ensuring a mutual agreement between all parties. If the overall product scope is not accurately described and agreed upon, conflict and rework is almost certain to occur.

3.5.1.1 Finalizing Needs Analysis

Stakeholders have expectations that drive needs, which constitute the building blocks of the project and product scope.

The development of the initial product scope is initiated following the stakeholder analysis and captures the wide range of expressed needs (see Figure 3.13). The product development team will collect facts and arrange for discussions and interviews with the appropriate stakeholders. The project manager must seek agreement from all parties on the contents of the collected needs, as this will subsequently enhance buy-in from all stakeholders.

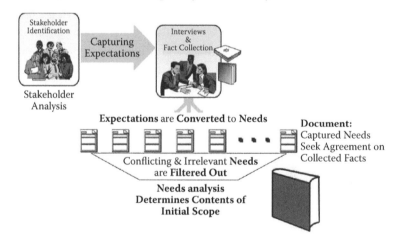

FIGURE 3.13
Needs analysis—stakeholder collection

3.5.1.2 Developing Requirements

The product requirements correspond to the functions that the product must provide to satisfy the agreed-upon needs. In requirements engineering, functional requirements specify particular results of a system.

A functional requirement defines a function of a system or its component. Functional requirements are processes and information manipulation and processing, including calculations, which define what a system is supposed to accomplish. Functional requirements are supported by nonfunctional requirements (also known as quality requirements), which impose constraints on the design or implementation (such as performance requirements, security, or reliability). How a system implements functional requirements is detailed in the system design.

Functional requirements drive the application architecture of a system, whereas nonfunctional requirements drive the technical architecture of a system.

The project manager must establish that all needs have been addressed by the requirements. All needs that have been deemed to be unnecessary, superfluous, or unfeasible, are to be recorded and the corresponding requesting stakeholder informed that the original expectation will not be within the scope of this project. Approval of the product requirements document must be sought by the project manager before proceeding with the product design.

3.5.1.3 Finalizing the Design

The product or systems design is the process of defining the architecture, components, modules, interfaces, and information for a system to satisfy specified requirements. Designers may produce one or more "models" of a system for review with stakeholders and clients.

The major challenge for the project manager is to establish with the design team the appropriate level of detail for the solution design. This detail will have an impact on the accuracy of the estimations and the assumptions to be established.

The solution design must be reviewed for coherence to the product requirements. Functional requirements that have not been covered and their motives are documented and the corresponding stakeholders informed. Approval of the product design document must be sought by the project manager before finalizing the project plan.

3.5.2 Project Scope Development

The purpose of project scope development is to use additional knowledge about the product of the project to extend and refine the contents of project scope, which will provide more granularity in the WBS and detail the work to perform. A lower level of detail of work packages enables greater accuracy in estimating their duration and cost (in accordance with the human, material, and equipment resource mix chosen) and allows the project team to ascertain with more confidence the dependencies among project activities, and enhance the quality of the project schedule. (The schedule will subsequently be adjusted according to the availability of resources.)

A detailed project scope improves the understanding and definition of the processes and standards that will be used to measure quality during project execution and control.

When developing the project scope, the project manager should create a revised version of each document while maintaining the integrity of the original documents. This will provide an audit trail as to how the project scope has evolved throughout the project life cycle.

As was described previously, project scope steps through a series of qualifications, which describe its level of detail:

- **Scope statement**—describes the major objectives and states the project deliverables
- **Scope description**—describes the needs, main theme, and key components of the project
- **Scope definition**—breakdown of the project corresponding to the functional requirements
- **Scope specification**—comprehensive design of the project to be delivered

Thus it is to be expected that the project scope, and the project's knowledge base, will evolve in detail, contents, and volume. The volume of changes to the project scope is directly related to the breakdown of the project to its design and subsequent realization, the unknowns identified, the assumptions made, the decisions that can only be made following potential future events, and the volatility of the project environment.

The title of the movie may not change; however, the storyboard and the script will.

3.5.3 Managing Unknowns in the Project Scope

Throughout the development of the project scope, the project manager and the core team will address the challenge of how to manage unknowns. Moreover, at all stages of the project development life cycle, unknowns will exist until final close-out and handover. This may seem odd to many; however, it is obvious that at the start of project scope development, among many other things, the project schedule is not known, since the WBS, estimation, and network of activities have not as yet been developed. The question to the project team is how to determine how to know what they do not know—the unknowns. (Aristotle is quoted to have said, "Wisdom comes when I know that I do not know.")

Establishing unknowns for a project is not as mysterious as it sounds. Figure 3.14 presents a scheme for establishing a list of unknowns. As the project scope development proceeds, the project team maintains a record of what is known about the project and what is still to be known or discovered—there are the unknowns. The knowns represent the real and validated facts about the project. This is the current project documentation. Establishing unknowns is done by determining three different categories of information that the project does not possess as yet: missing data, unqualified statements, and language issues.

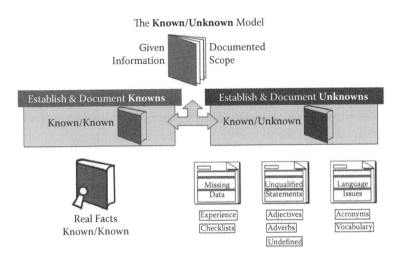

FIGURE 3.14

The known-unknown model

3.5.3.1 Missing Data

From the documentation on hand, a list of missing data can be drafted when the experienced reader cannot find certain essential information needed to pursue the project. For example are simple things like organizational charts known; are volumes of transactional data known; do floor plans exist, etc. For the inexperienced reader, or as assistance to those experienced, a checklist can be used that identifies those items and elements of needed information.

3.5.3.2 Unqualified Statements

Most if not of all documentation is fraught with unqualified statements (this book included). Drafting a list of unqualified statements is primarily a grammar exercise of pinpointing all statements that contain adjectives and adverbs, for example, a robust system, an easy-to-use procedure, or rapid access to parking facilities. All these statements are not quantifiable. To the list can be added all classical "to be defined" statements that abound in documentation. Many are stated as TBD. Others may be stated as "this will be addressed at a later stage." The worst is finding a totally blank page in the documentation with a statement, "This page has been left intentionally blank, awaiting further information."

3.5.3.3 Language Issues

The third category covers comprehension and understanding of the written documentation. Without a glossary of terms or a lexicon, all 2-, 3-, 4-, or even 5-letter acronyms are the biggest contenders to be added to the list of unknowns. In specialized documents, all misunderstood or newly discovered esoteric vocabulary to the reader should also be added to the list. The list must also include any foreign or strange words unknown to the reader. Furthermore, as all who write any documentation are not expected to hold a masters degree in English language, the unknown list can be expanded to include whatever portion of the document that just cannot be understood by the reader(s).

3.5.3.4 Resolving and Answering Unknowns

From the established list of unknowns, the project team proceeds to answer these and seek information to convert the unknown to a known. Basically, performing further analysis, conducting interviews, collecting data, and any other activity that searches for the required information. Figure 3.15 presents a scheme for addressing unknowns: **Get, Assume, and Park—GAP.**

Managing **Unknowns** of the **Project Scope**

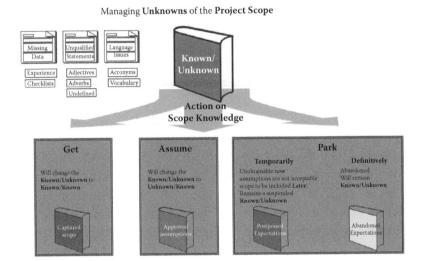

FIGURE 3.15
GAP analysis for known-unknown

3.5.3.5.1

Get   This is the most common of actions. The project team will seek to retrieve information and the knowledge by asking for it, searching for it or performing work to obtain it. For example, requesting the organizational chart or going on-site and measuring the floors to draft the floor plans. The newly obtained knowledge then shifts from an unknown state to one of being known and documented. Thus the project scope knowledge base increases in detail and volume.

3.5.3.5.2

Assume   However much or hard the project team searches for unknown information, there are many instances when the unknown remains in its current state. This occurs usually when the team is faced with either a tight timeline, when there is not the material time to get the information, or the information is inaccessible, for reasons of confidentiality or flat refusal and rebuke from the source.

In many cases, the unknowns address information that cannot be known at this moment of time. For example, the volume of today's data transactions is known; however, the evolution of this data over the period of the project development and beyond is not known. Similarly, this is the case for all information that can only be known at a future date, such as new legislation or rate of inflation.

The project team can now only proceed by establishing assumptions. These must be acceptable, realistic, coherent, feasible, and most of all approved by all the parties involved in the assumption. All assumptions must be recorded clearly and documented thoroughly, as they will constitute part of the project scope, albeit as "virtual" knowledge.

3.5.3.5.3

Park The remaining unknowns from the original list are then assessed and analyzed, and the project team decides on the appropriate actions to resolve these. Many of the unknowns are latent from the Assume step, as no valid or acceptable assumption can be made. The project team then has to make a choice of either including the unknown item to be considered within the project, but at a future date, or simply place it out of the scope of the project.

The team will Park the unknown either as a suspended scope item or as an out-of-scope item.

The suspended scope item means that the expectations related to the unknown cannot be descoped, nor can assumptions be made with the current knowledge on hand. These suspended scope items will be reviewed at a later stage in the scope development when the information becomes available or a valid assumption can be made. This is documented and a date set for the review of the unknown. For example, training has to be provided for new staff to be hired. Based on today's information, the HR department cannot give information on the volume or the skills/competence of staff to be hired. Assumptions may be of no value at this stage, and the item cannot be descoped. A date is set with HR, when the training provision will be again addressed within the project scope.

An out-of-scope item means that the project will no longer consider this expectation as part of the project scope. For example, there may be a need to expand the car park facility of the new offices, but this depends on the company's growth, the future economic climate, and whatever new legislation may be drafted by the authorities. All out-of-scope expectations are documented and approved.

3.5.4 Constituents of the Project Scope

As has been discussed, project scope development is a continuous process during which the project team addresses the product scope contents while maintaining the balance between knowns and unknowns. As work

Key Documents - What is in : What is Out

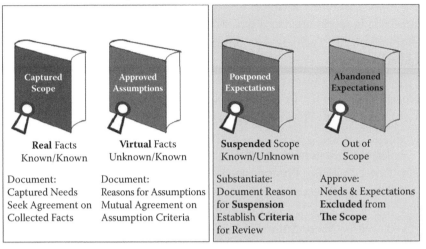

Captured Scope	**Approved Assumptions**	**Postponed Expectations**	**Abandoned Expectations**
Real Facts	**Virtual** Facts	**Suspended** Scope	Out of
Known/Known	Unknown/Known	Known/Unknown	Scope
Document:	Document:	Substantiate:	Approve:
Captured Needs	Reasons for Assumptions	Document Reason	Needs & Expectations
Seek Agreement on	Mutual Agreement on	for **Suspension**	**Excluded** from
Collected Facts	Assumption Criteria	Establish **Criteria**	**The Scope**
		for Review	

FIGURE 3.16
Key scope documents

progresses, the project manager ensures that the project documentation reflects what is *in* the scope and what is *out* of the scope. The four major documents in Figure 3.16 demonstrate this clearly.

3.6 THE WORK BREAKDOWN STRUCTURE

The project work breakdown structure (WBS) is exactly what it says: It is the structure that breaks down the work required to complete the project.

The WBS is a deliverable-oriented hierarchical decomposition of the work to be executed by the project management team to accomplish the project objectives and create the required deliverables. It organizes and defines the total scope of the project. Each descending level represents an increasingly detailed definition of the project work. The WBS is decomposed into work packages. The deliverable orientation of the hierarchy includes both internal and external deliverables (*PMBOK® Guide*, p. 116).

3.6.1 WBS Levels

The WBS is represented as a tree structure, which shows a subdivision of effort required to achieve an objective; for example, a program, project, and contract.

The WBS is organized around the primary deliverables/products of the project instead of the work activities needed to produce the products.

The structure in Figure 3.17 is a common way to present the WBS:

- Level 1—System or Program.
- Level 2—Major Program Deliverables.
- Level 3—Project Level.
- Level 4—Major Project Deliverables.
- Level 5—Major Summary Activity Level: represents the minimum level of detail required to plan, schedule, and manage a project.
- Level 6 and if needed Levels 7 and 8—Work Package Levels contain task and activities when Level 5 does not contain sufficient detail to plan, schedule, and manage the work. Some Level 6 activities may need further breakdown to Level 7, and may eventually require a Level 8.

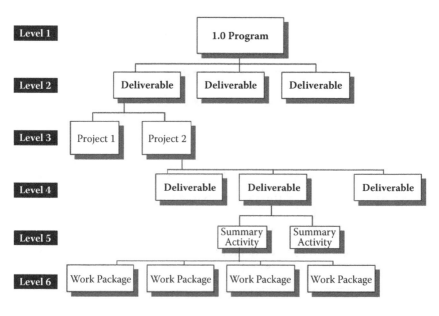

FIGURE 3.17
WBS generic structure

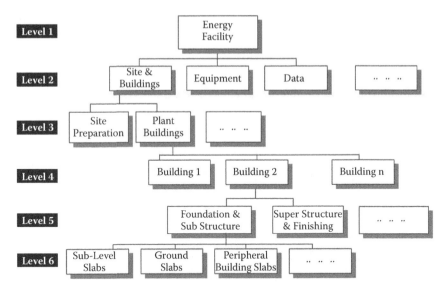

FIGURE 3.18
WBS example—energy facility

Since the deliverables are the desired results of the project, they form a relatively stable set of categories in which the duration and costs of the activities needed to achieve them can be collected. A well-designed WBS makes it easy to assign each project activity to one and only one terminal element of the WBS—a work package. In addition to its function in duration and cost estimation, the WBS also helps map functions from one level of system specification to another, for example requirements across documents (an example WBS is given in Figure 3.18).

A WBS permits the aggregation of subordinate costs for activities, materials, etc., into their successively higher-level "parent" activities, materials, etc. At this stage, activity durations cannot be aggregated, as this does not make any sense until a precedence analysis is performed and a logical network of activities constructed.

For each element of the WBS, a description of the activity to be performed is generated. This technique, called a WBS dictionary in PMI vocabulary, is used to define and organize the work to be performed for the project scope.

The initial WBS is not an exhaustive list of work, as many activities to be performed in the latter stages of the project cannot be broken down to their detail until a further date. The WBS should be treated as a comprehensive classification of project scope.

Note, too, that the WBS is not an organizational hierarchy. While it is common for responsibility to be assigned to organizational elements, and this is often reflected at the higher levels of the WBS, a WBS that maps exactly to the organizational structure is not descriptive of the project scope and is not deliverable-oriented.

A WBS is neither a project plan, nor a schedule, nor a chronological listing. A WBS is not a logic model, nor is it a strategy map. It is not a recommended practice to construct a project schedule (e.g., using project management software) before designing a comprehensive WBS. This would be similar to scheduling the construction activities prior to completing the design. The WBS hierarchy concentrates on deliverables and the required activities to produce them.

Updates to the WBS, other than progressive elaboration of details, require formal change control. As the WBS should be deliverable-oriented, changes in planned outcomes require a higher degree of formality.

One of the most important Work Breakdown Structure design principles is called the 100 percent rule.

The WBS includes 100 percent of the work defined by the project scope and captures all deliverables—internal, external, and interim—in terms of the work to be completed, including project management. The WBS should not include any work that falls outside the actual scope of the project, that is, it cannot include more than 100 percent of the work. It is important to remember that the 100 percent rule also applies to the activity level. The work represented by the activities in each work package must add up to 100 percent of the work necessary to complete the work package.

The 100 percent rule is one of the most important principles guiding the development, decomposition, and evaluation of the WBS. The rule applies at all levels within the hierarchy: The sum of the work at the "child" level must equal 100 percent of the work represented by the "parent."

3.6.2 Team Dynamics in WBS

The WBS is constructed by the project management core team. The project manager provides the framework, the known project scope, the project deliverables, and the identified assumptions and constraints.

The project manager will plan to engage the project team to analyze and break down the work into a sequence of deliverables, activities, and tasks until the detail defines a manageable project.

The WBS is structured such as to assign areas of responsibilities to members of the core team. Agreement is reached on how best to partition the WBS and the allocation of the corresponding deliverables. The core team can then proceed together to develop the WBS in a planning session, or each core team member can "take away" the assigned project segment and further develop the segment's WBS with members of a subproject team. In this case, the core team then meets to collate and consolidate the different segments to complete the overall project's WBS.

The main objective of this distribution of work is to assign portions of the WBS development to those who have the best expertise and experience, and who can develop a detailed structure. Obviously it goes without saying that team member participation greatly enhances ownership and commitment.

3.6.3 Seeking Granularity of Work Packages

The challenge of constructing a WBS is to determine the level of detail, the granularity of the tree structure, which best describes the comprehensive view of the needed work activities, while allowing for subsequent acceptable duration and cost estimations. A WBS can be compared to a tree, with its trunk, branches, and twigs. The work packages are the leaves, and it is there that the work is performed.

The use of standard WBS templates or those from previous similar projects is strongly suggested, as it will avoid "reinventing the wheel" and allow the project team to focus on the specificity of the project scope that differs from a given standard WBS.

As stated previously, the higher levels of the WBS refer to each deliverable described in the project scope. The WBS can also start with a functional component of the scope (foundations, walls, roof, etc.), a competence or a discipline (hardware, software, contracts, etc.), the phases of the project development cycle (requirements, design, delivery, etc.), or even a geographic breakdown (headquarters, office A, on-site, etc.). Whatever the chosen structure, the highest level represents the principal segments of the project and will correspond to the assignment to core team members.

The higher levels of the WBS will more often than not be defined as objects or nouns. Another technique can be used by starting with major verbs, as in the case of project phases.

Each block of the WBS is described as a statement that can be broken down into smaller, more manageable components (trunk, branch, twig, leaf). This is repeated until the components are small enough to be defined in the greatest possible detail. As this stage, the WBS elements will be described as sentences with action verbs. The tree structure is now approaching or is at the "leaf"—work package level. Each branch and twig can be expanded to different depths, depending on the nature of the work and the understanding the team has of its contents. The WBS decomposition exercise assists project team members to better understand and properly document the project scope. It also provides information needed for project estimation and scheduling. The breakdown can be judged to be at a granular level when:

- The work package can be clearly defined and is restricted to one or two simple sentences.
- The activity result/deliverable can be reduced to one output only.
- There are few assumptions, and they are admissible.
- The mix of resources needed to perform the work are kept to a minimum.
- Work package duration can be realistically and confidently estimated with little inaccuracy.
- Costs can be assigned without the need to break down further.
- It is a unique package of work that can be outsourced or contracted out.
- It makes no sense to break down any further.

It is important that those who develop the WBS feel comfortable with the level of granularity reached. This will condition the subsequent activity estimations and construction of the activity network.

3.6.4 Techniques for Building the WBS

The WBS can take multiple forms, and each format has its own advantages. The two most common forms for a WBS are graphic and tabular (indented). The same information can be displayed in different formats.

The project manager proposes the top level(s) of the WBS. Core team members review the suggested WBS and define the next level of detail. Work is collaborative to identify subsequent levels by drafting WBS activities and tasks. A very dynamic and engaging way is to develop the WBS with Post-its® and place these on a wall.

The draft WBS is published to team members and others to verify and check the contents, and identify additional work.

In many engineering projects, a product breakdown structure (PBS) or bill of materials (BOM) can be used to guide the WBS build process.

3.6.5 Different Representations of the WBS

3.6.5.1 Graphical

This is the most powerful representation technique, as it guides the breakdown, and team members can develop multiple branches simultaneously. It also offers the possibility, especially when using Post-its as a support, to move or shift around the identified work activities so as to achieve greater coherence in the tree structure.

When using a standard template, there is often a temptation to skip reconstructing the WBS as a team effort. However much this could seem a somewhat tedious step, the benefit derived from engaging the team members and increasing their collective knowledge greatly outweighs what could be considered time wasted.

Two main techniques can be used to build the WBS: top-down or bottom-up (see Figure 3.19).

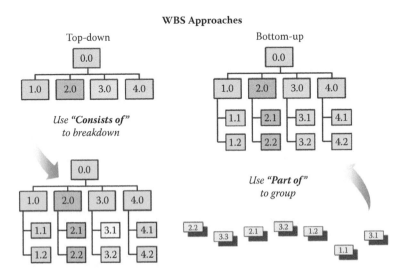

FIGURE 3.19
WBS approaches

Top-down consists, as the words imply, to work from level 1 and progressively breaking down through the lower levels. This is particularly effective when the team possesses a WBS template and/or is sufficiently conversant in the main components and can develop the structure by "drilling" down to the detail. The technique follows a classical breakdown method of assessing each developed activity in the WBS and asking the question, "What does this consist of?" It should be noted however, that the top-down approach conditions the structure once levels 1 and 2 have been established.

The bottom-up approach is predominantly used to structure a new project when no previous templates are available or similar previous project historical data exists. The approach is also often used to engage the team members and enhance their participation and thus their ownership of the WBS. The technique calls upon brainstorming, in which each team member develops as many activities as they can, irrespective of level. All activities, written on Post-its, are placed on the wall. The team members then structure the WBS upwards by using affinity techniques and asking of each activity "What is this part of?"

3.6.5.2 Tabular

This technique is an extension of a to-do list. Many people prefer this technique when dealing with a short list of activities and when working alone. However, it may reduce team member engagement, as the manner to develop the list of work activities is often performed by one person, even though others may participate vocally. The technique has a disadvantage, compared to the graphic WBS, in that it is at times cumbersome to modify or expand intermediate levels.

The resulting structure resembles a structured table of contents as presented here:

> Formalize solution/product
>   Concept exploration
>   Initiate start of phase
>   Develop identification report
>   Search literature
>   Interview involved people
>   Define purpose, scope, objectives, and constraints
>   Prepare identification report
>   Perform needs analysis

> > Collate/gather enabling information/documents
> > Scope solution and agree terms of reference
> > Plan for initial study
> > Information/data gathering & briefing
> > Modelling the business/market environment
> > Plan for business analysis meetings
> > Conduct business analysis meetings
> > Review and consolidate business analysis information
> > Complete business analysis report
> > Produce total system arch. And general alternatives
> > Produce strategic solution development plan
> > Prepare management reports and presentation scenario
> > Management presentation
> > Obtain management approvals and concurrences
> Close off end of phase

Requirements development
> Initiative start of phase
> Gather information/data
> > Study present and proposed systems
> > Schedule interviews
> > Conduct and document interviews
> Describe present system
> > Observe present system
> > Analyze performance of current system
> > Prepare written description of present system
> Develop system requirements specification report
> > Detail purpose, scope, objectives of proposed system
> > Define system solution—top level
> > Determine system impact
> > Establish performance requirements
> > Define performance requirements
> > Produce installation/conversion strategic plans
> Complete solution requirement plan
> > Prepare initial training plan
> > Produce project/product ctrl procedures/documents
> > Specify information recovery requirements
> > Review/agree and approve system design requirements
> > Produce project plan for solution development
Close off end of phase

3.6.5.3 Numbering Systems

Whatever technique is used to build the WBS, a numbering system should be used. This is, of course, to refer uniquely and specifically to any one item in the structure. Furthermore the numbering system will also allow identifying quickly the level of the WBS item and can also provide for cost center assignment.

1	02			REQUIREMENTS DEVELOPMENT
1	02	01		INITIATIVE START OF PHASE
1	02	02		GATHER INFORMATION/DATA
1	02	02	01	STUDY PRESENT AND PROPOSED SYSTEMS
1	02	02	02	SCHEDULE INTERVIEWS
1	02	02	03	CONDUCT AND DOCUMENT INTERVIEWS
1	02	03		DESCRIBE PRESENT SYSTEM
1	02	03	01	OBSERVE PRESENT SYSTEM
1	02	03	02	ANALYZE PERFORMANCE OF PRESENT SYSTEM
1	02	03	03	PREPARE WRITTEN DESCRIPTION OF PRESENT SYSTEM
1	02	04		DEVELOP SYSTEM REQUIREMENTS SPECIFICATION REPORT
1	02	04	01	DETAIL PURPOSE, SCOPE, OBJECTIVES OF PROPOSED SYSTEM
1	02	04	02	DEFINE SYSTEM SOLUTION—TOP LEVEL
1	02	04	03	DETERMINE SYSTEM IMPACT
1	02	04	04	ESTABLISH PERFORMANCE REQUIREMENTS
1	02	04	05	DEFINE PERFORMANCE REQUIREMENTS
1	02	04	06	PRODUCE INSTALLATION/CONVERSION STRATEGIC PLANS
1	02	05		COMPLETE SOLUTION REQUIREMENT PLAN
1	02	05	01	PREPARE INITIAL TRAINING PLAN
1	02	05	02	PRODUCE PROJECT/PRODUCT CTRL PROCEDURES/ DOCUMENTS
1	02	05	03	SPECIFY INFORMATION RECOVERY REQUIREMENTS
1	02	05	04	REVIEW/AGREE AND APPROVE SYSTEM DESIGN REQUIREMENTS
1	02	05	05	PRODUCE PROJECT PLAN FOR SOLUTION DEVELOPMENT

3.6.6 Case Study: Develop the Project's WBS

Workshop 3: Project Strategic High-Level Plan

Today is Thursday, 14th November.

Management has approved the project's objectives, and especially the financial estimates for subproject LZB-Sales. Your next meeting with management is scheduled for the 27th November, during which you will present the detailed LEY-WORLD project plan and cost/schedule.

You are entering the project planning stage.

Establish the initial project structure for project LEY-WORLD:

- The high-level WBS—Work Breakdown Structure
- The project's organizational breakdown structure, representing the individuals responsible for each subproject.

3.7 ESTIMATING

Estimation is an inaccurate quantification of duration and cost of an activity to be performed at a future time with unsecured resources under uncertain conditions in an environment that is not under the project's control. The range of inaccuracy differs throughout the project planning and execution phases—the range will narrow. However, project managers must accept that estimates are approximations and incorrect and must convey this to sponsors, management, clients, and key stakeholders. Powerful tools and extensive databases can provide highly accurate results; however, these will never be without an error rate. The key is to know by how much they are off the mark and to narrow the range as the project progresses.

Establish the duration and the cost of a project based upon insufficient data, often ambiguous and incomplete, by determining the most probable cost and schedule of an activity performed by another party, over which one has little of no possibility for direct influence, in a future time frame, which is not yet determined!

Project managers must accept that estimates are not self-fulfilling prophecies and that they are "managing inaccuracies in a world of uncertainties."

The financial estimating focus is the *cost* of the project to the performing organization. This is not to be confused with determining the *price* of a commercial project to a client purchasing/buying organization.

An additional key point to remember is that estimating is not an exact science. Whatever tools and techniques are used, and however much data can be used from previous projects, the project manager must focus on the range of inaccuracy of the estimate.

There are also certain behavioral issues that need to be considered in estimating.

- Different perspectives exist, based on the managerial level.
- Senior people tend to underestimate, junior people tend to overestimate.
- Lower organizational levels tend to arbitrarily add reserves, upper levels to arbitrarily delete them.

3.7.1 Underlying Principles

The Engineering V, discussed in this and previous chapters, represents a structured project development life cycle chronology, and how it guides the project from the initial needs to the delivered solution. The major steps and phases in the "conceptualize" segment will produce more detail to the problem and the proposed solution. We also previously introduced the knowledge base and how, during the problem resolution exercise of the "conceptualize" portion of the project, the project manager and the team members will define, describe, collect, and collate the documentation that reflects the current state of the project's scope. Throughout the "materialize" portion of the project, more detail will be attained and more documentation will be developed.

Estimates are necessary at all times throughout the project (see Figure 3.20). It can be very easily understood that those estimates developed at the early stages of the project cannot be other than gross approximations that can only give a general idea of both duration and cost. Estimates will also be required during the latter stages of project execution leading up to project close-out and handover.

Estimating project work is a project management core competency. Project managers should be familiar with, and be able to use or direct the project team, to use the appropriate tools to ensure that estimates provided during the planning process are reliable.

There are a number of factors that should be considered during the estimating process. Every project is different regardless of how similar it may seem to other projects that have been performed in the past. Failure to recognize the following factors may well lead to an unrealistic and inapplicable estimate:

- Misinterpretation of the statement of work
- Omissions to the scope
- Incorrect assumptions
- An excessively optimistic schedule

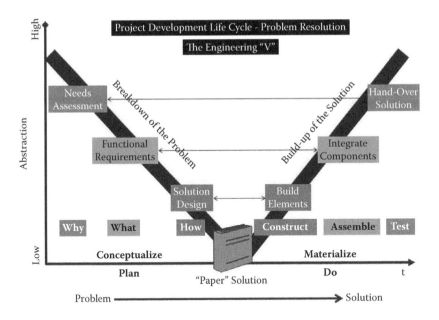

FIGURE 3.20
Estimating underlying principles

- Incorrect skill levels applied to project activities
- Failure to account for risks
- Failure to account for cost escalations such as contractual labor rate increases and inflation

3.7.1.1 Understanding Estimating Inaccuracy

An estimate is a prediction of the expected final duration/cost of a proposed project, for a given scope of work. By its nature, an estimate is associated with uncertainty, and therefore, is also associated with a probability of over/under-running the predicted duration/cost.

Prediction with 100 percent accuracy does not exist. Duration/cost estimates have to be stated in ranges or probability of not being exceeded. Variances to the estimate will be known during project execution progress monitoring, when actual performance is recorded. When an estimate is found to vary from actual performance, the reasons for the variance should be identified, and the lessons learned should be incorporated in future estimates to improve their accuracy.

There are many levels of estimating accuracy—from initial order of magnitude to detailed definitive.

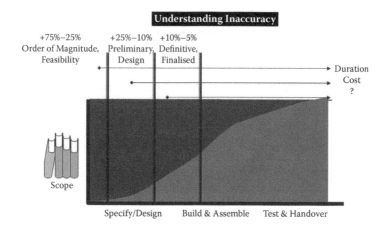

FIGURE 3.21
Understanding inaccuracy

The project manager should be clear about what kind of estimate can be given at the corresponding stage or phase of the project and must clearly communicate the accuracy range being provided. The industry-recognized table (Figure 3.21) indicates the types of estimates and their corresponding estimating ranges. For certain mature industries, where there is a sound and rich history of previous and similar projects, the ranges would be narrower, especially at the feasibility and preliminary design stages.

The detailed and definitive project estimate, together with a correspondingly detailed scope of work, and providing the highest accuracy, can only be constructed from a matching detailed WBS.

3.7.1.2 Incorporating Assumptions and Risks

Estimation is an inaccurate quantification of duration and cost of an activity to be performed at a future time with unsecured resources under uncertain conditions in an environment that is not under the project's control. Thus estimates are fraught with assumptions. These assumptions have to be well described and validated as being realistic.

Assumptions are critical to estimating. All estimates are based on a set of assumptions regarding the nature of the product, the resources and their availability, tools and technical approach, and other factors. Always communicate assumptions along with the estimate. If assumptions change, the estimate will change.

In addition to describing the basic duration/cost parameters used in developing the estimate, the basis must include descriptions and supporting data for all assumptions used and qualification to the data used in the estimate. Assumptions may be grouped by category, if appropriate for specific projects. In general assumption categories would cover:

- **Design Assumptions**—facility, process, equipment and material assumptions affecting the duration/cost estimate. Design assumptions may include such items as: technology, specific safety and environmental needs, and regulatory requirements
- **Economic Assumptions**—resource unit rates, resource productivity, equipment and material pricing, taxes, exchange rates, and other monetary impacts on the project
- **Logistics and Execution Assumptions**—purchasing and contracting philosophies, resources availability, special transportation requirements and other logistical requirements
- **Schedule Assumptions**—Equipment and material lead times, fabrication and construction planning, weather impacts, and interrelationship between tasks and interrelated projects.

Risk assessment is an integral part of the estimating process. It is used to identify reserves, identify the degree of expected variance from the plan, and give stakeholders a sense of the accuracy of estimates.

There is temptation at this stage of estimation to address the risks associated to an activity by applying "padding" or even concocting a multiplication coefficient to allow for "more money" or "more time" to perform the activity. Care must be taken that additional duration or cost that have been applied to respond to risks, be clearly explained and substantiated. Otherwise, during duration/cost negotiations with management or client these disappear into thin air, leaving no trace, and position the project activities back to their initial uncertainty. It is strongly suggested to avoid the "padding" technique and to conduct a risk assessment and planning. This will be presented in Chapter 5, Risk Management.

3.7.1.3 Differentiating between Estimating and Costing

Estimation is the calculated approximation of a result that is usable even if input data may be incomplete or uncertain. In mathematics, approximation or estimation typically means finding upper or lower bounds of a quantity that cannot readily be computed precisely and is also an educated guess.

Costing, for the purposes of this chapter, will be considered as the directly incurred costs for an element or an item. These are costs that are explicitly identifiable as arising from the conduct of a project, are charged as the cash value actually spent, and are supported by an auditable record; for example, a flat rental fee is a directly incurred cost; a catalogued product in a price list is a directly incurred cost; an airplane ticket is a directly incurred cost, etc.

Thus a fully developed project cost will be a mix of inaccurate estimates and precise costs.

3.7.1.4 Assessment of Resources That Drive Duration and Cost

There are three categories of resources in projects that drive duration and cost: human, material, and equipment. These constitute the major direct costs to the project. Other direct costs can be categorized under services and include travel expenses, living accommodations for team members, transportation and customs, legal fees, fees for rentals, and other elements that are directly incurred costs.

The project manager must also consider the indirect costs that will be charged to the project. These are the overheads that the corporation's financial division will assign to each project. Each organization will have a different mechanism for the cost allocation of management, shared services, office rental, etc.

During estimation of the work packages, it is imperative that resources are identified by their characteristics—a *resource profile* (see Section 3.7.4, Identifying Resource Profiles). There is strong temptation to estimate for, and assign, a named resource to perform a work package. This is strongly not recommended, since at time of estimation there is no way to determine when that named resource will be required nor its availability, as the logic network and primary schedule have not yet been developed. Estimation imposes that the project manager and the team members consider different generic resource profiles, using an unlimited and unconstrained resource model. Once the primary schedule is developed and the resource loading per resource profile is established (see Chapter 4, Project Scheduling), then staffing for human resources can be initiated, and procurement for material and equipment resources can be performed. It is then that the unlimited and unconstrained resource estimation is confronted by the realities of limited and constrained resource availability.

- **Human resources:** refers to the skills and competence required of persons who will participate in any way in the performance of work on the project.
- **Material resources:** refers to those items that will be consumed in the performance of a work package, such as liquids, solids, electrical or mechanical components, energy, piping, cables, sand, etc. Estimation identifies quality and quantities, volumes, lengths, weights, etc.
- **Equipment resources:** refers to those items that will be utilized through hire, purchase or borrowing, to facilitate the performance of a work package. Items such as vehicles, tools and machinery, computers and software, etc., are equipment resources. Estimation identifies capacity and quantities, etc.

When an equipment resource is purchased, capital ownership and amortization must be calculated and the associated costs included in the estimate (this is usually determined by the financial division).

Estimation of duration and cost must consider the resource mix required to successfully perform the work package (see Section 3.7.5, Understanding the Nature of the Work to be Estimated).

3.7.2 Selecting the Work Packages to Be Estimated

As seen previously, the WBS is the structured chart that provides for work decomposition to arrive at the appropriate work package detail level desired for the project. Estimation is made at the work package level. The upper or parent levels can then be aggregated only for costs. The project manager and the core team members select the work package (the leaves) from the WBS, and then proceed to estimate each.

As was noted previously, the descriptions of each work package have been developed and collated in a WBS dictionary (see Figure 3.22). Figure 3.23 is an example format for a work package in the WBS dictionary. As can be seen, before proceeding to estimate, each work

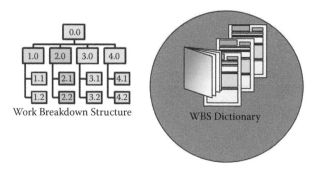

Work Breakdown Structure   WBS Dictionary

FIGURE 3.22
WBS dictionary

package would have a comprehensive description of the work to perform, the deliverables it will produce, and the assumptions made. At this stage, the estimation can proceed to identify the resources required and subsequently determine the estimated duration and cost of the work package.

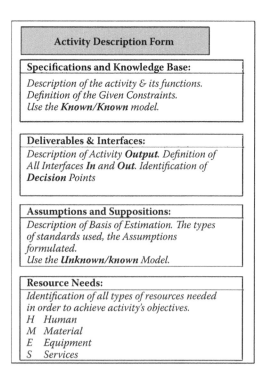

FIGURE 3.23
Activity description form

3.7.3 Time Units for Duration Estimation

Before estimating the duration of a work package, the project manager must establish with the project team the method and unit to use to determine the work load:

- Human resources provide effort, and this will be in a time unit of hours or days. It may even be minutes or weeks.
- Material resources are consumed, and this is usually expressed as a quantity, a volume, a length, or any other unit.
- Equipment resources are utilized in time units.

To this list, other resources used such as facilities are to be considered. The duration estimation will be driven by the resource that will be the most used by the work package. For example, a machine test requires an operator to start and stop the test, but it is the machine that dictates the duration of the activity.

Human and equipment resources' work loads are estimated in time units, and this estimate needs to be converted into "working time"—a period of "legal" time for human resources during which actual work on the work package can be performed, or a period of "usage" time for equipment resources. "Elapsed time" (calendar time) is not estimated at this stage, as this can only be done once the network of activities is constructed and the schedule produced against a civil and corporate calendar.

The project manager must also establish the standard productivity (rate at which work is performed) for each resource profile and the standard availability (resource accessibility). It is incorrect to state that knowledge workers who have a legal work day of eight hours will have a 100 percent standard productivity of eight hours, as nonproductive time such as meetings must be deducted. Similarly, when determining equipment standard productivity, the project manager must consider the portion of nonproductive downtime, which will reduce the supposed 100 percent rate.

As will be explained in Section 3.7.5.1, Linear Productivity, the productivity rate and the availability are key parameters when determining work package duration.

3.7.4 Identifying Resource Profiles

3.7.4.1 Human Resources

This refers to the skills and competence required of those who will participate in any way in the performance of work on the project. This will cover all skill profiles required from low-skilled labor to highly specialized knowledge workers. The project will "exploit" the skills and competences of the human resources (see Figure 3.24).

The human resource profiles clearly define the skill and competence characteristics, the expected or assumed productivity rate of the resource, and the unit cost of the resource in an appropriate time measure: work hours, work days, or longer.

In establishing a standard productivity rate for a human resource profile, the project manager should take into account absenteeism, meetings, discussions, and staff interaction to determine the real net daily productivity availability for human resources. Future resources to be assigned may have ongoing operational responsibilities occupying a portion of their time, and the average linear nonproductive percentage of legal hours must be factored into the schedule. Once effort estimates have been determined for each activity, the project schedule must be revised to reflect them.

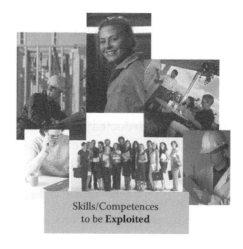

Skills/Competences
to be **Exploited**

FIGURE 3.24
Human resources—Skills-competences to be exploited

Quantities & Volumes
to be **Consumed**

FIGURE 3.25
Material resources—quantities and volumes to be consumed

3.7.4.2 Material Resources

This refers to those items that will be consumed in the performance of a work package, such as liquids, solids, electrical or mechanical components, energy, piping, cables, sand, etc. Estimation identifies quality and quantities, volumes, lengths, weights, etc. (See Figure 3.25.)

The unit cost of the material resource is to be recorded per each profile. This is usually done from the project bill of materials and/or parts list.

3.7.4.3 Equipment Resources

This refers to those items that will be utilized in the performance of a work package either through hire, purchase, or borrowing. Vehicles, tools and machinery, computers and software, etc., are equipment resources. Estimation identifies capacity and quantities, etc. (See Figure 3.26.)

The equipment resource profile must state clearly the expected or assumed productivity usability of the resource, and the unit cost of the resource in an appropriate time measure: work hours, work days, or longer.

3.7.5 Understanding the Nature of the Work to Be Estimated

Each work package identified describes the nature of the work and its contents. Estimation aims to determine first the work load of the resources required to perform the task. The work load estimate for human resources provides the effort, in hours, days, or any other established time unit. The estimate for material resources will indicate the quantities,

FIGURE 3.26
Equipment resources—capacities and quantities to be utilized

volumes, or measures in a corresponding unit to the resource: numeric, liters/gallons, meters/feet. Equipment resources estimates provide for usage over a time period, in hours, days, or weeks. From the work load and the estimated units to use, the cost can be determined by applying the unit cost to the work load.

The project manager must possess data on the productivity full-time equivalent of the identified resource profiles, as discussed in Section 3.7.4, Identifying Resource Profiles.

The nature of work packages will fall principally into two major categories: linear or variable productivity. The former refers to work that calls upon resources that have a continuous or constant rate of productivity, where duration and cost can be determined by converting directly the required effort or work rate into working time—this would be the case for low-skilled labor and equipment. The latter refers to work that utilizes resources whose productivity depends more on the content of the work package. No direct conversion is possible and the working time must be estimated—this would be particularly the case for skilled knowledge resources.

3.7.5.1 Linear Productivity

As stated previously, linear productivity refers to applying a continuous or constant rate of productivity to the work performance. The determination of duration and cost is a function of effort, consumption, and/or usage against working time and unit cost of the resource.

The key parameters used are the productivity rate and the availability:

Duration in Work Time = [(EFFORT or USAGE) / PRODUCTIVITY] / AVAILABILITY

Cost = [(EFFORT or USAGE) / PRODUCTIVITY] × Unit Cost

Example for Duration

Human Resource HR01 is to perform 50 hours of effort and has a productivity rate of 80 percent and an Availability of 50 percent.

Duration in Work Time
= [(50)/0.80]/0.50
= 62.5/0.50
= 125 hours
If the legal work day is 8 hours a day, then the duration is
= 15.6 work days

Example for Cost:
Human Resource HR01 has a Unit Cost of $15
Cost in $
= [(50)/0.80] × 15
= 62.5 × 15
= $937.5

Modifying the work load or the productivity has a direct impact on the duration and cost of the work package.

3.7.5.2 Variable Productivity

When the work package estimate is driven by the nature of the work contents and requires a knowledge-worker human resource, a direct formula as given in the example cannot be used for the duration calculation. The project manager and the team members must then choose to use one the estimation tools described in the next section.

3.7.6 Estimating Techniques

The estimation techniques are designed to assist in the process of approximating the level of project resource effort/usage required, work time duration, and their associated costs.

Each method has strengths and weaknesses. When possible, estimation should be based on several methods. If these do not return approximately the same result, there is insufficient information available, and action should be taken to collect more detail.

In addition, a formal risk assessment is an essential prerequisite for project estimation. (See Chapter 5, Risk Management.)

3.7.6.1 Analogy

Analogous estimating, also known as top-down estimating, relies on previous or similar projects to establish a broad or high level estimate. This type of estimate relies on expert judgment, lessons learned, and usually requires adjustments throughout the project.

Analogy-based estimation is a technique for early life cycle macro-estimation. Analogy-based estimation involves selecting two or more completed projects that most closely match the characteristics of the planned project. The chosen projects, or analogues, are then used as the base for the new estimate. The steps below are usually followed to formulate the estimate:

- Establish the attributes of the planned project.
- Collect data for projects that closely match the attributes of the planned project.
- Use the known development effort from the selected projects to establish and use as an initial estimate for the target project.
- Compare each of the chosen attributes and adjust the initial effort estimate in light of the differences between the analogue and the planned project.

It is important to record the adjustments made and the set of assumptions governing the estimate.

This technique is an order of magnitude estimate, with a -25%:+75% range, and can be used for an early indication for deciding to pursue the project.

Analogy: Short Exercise

Installation of a CCTV system in a motorway gas station, to cover a filling area of 1500 m² with twelve pumps and one external payment kiosk, a 1200 m² one-level L-shaped building consisting of a central payment desk, a boutique with a cash desk, a snack-bar counter with an adjacent dining area, a beverage dispensing area, a rest area, washroom facilities, three administrative offices accessible only to staff, a storage room of 1000 m², and a 1500 m² car park.

Previous completed projects have equipped stations with a maximum of four pumps in a 500 m² area, with a rectangular 500 m² building consisting of a payment desk and a beverage dispensing area, basic washroom facilities, but with no boutique, snack-bar, or car park. A typical fully installed and tested system of three external cameras and six internal cameras cost $50,000 for a duration of one calendar month using a team of four experienced technicians. What analogies can be made?

3.7.6.2 Parametric

The parametric estimating technique utilizes multiplication models to calculate duration for linear productivity work packages and/or derive costs for consumption or usage of material and equipment resources. Similarly and when required, cost of services can be estimated using this technique.

Very often this technique is associated with the analogy estimation technique.

Short Exercise

Derived from previous projects and using a team of two experienced technicians, the installation and testing of one CCTV would require four hours of effort for each technician and half a day work time at a cost of $35 per technician hour, $2,500 for one CCTV, and 200 meters of cabling.

The new station will require twelve external and twenty internal CCTVs. Using two experienced technicians working a full-time, eight-hour working day, what would be the duration and cost of the installation and testing?

3.7.6.3 Subject Matter Experts

This technique is often used in the many situations in which historical data is either not available or has not been recorded in a usable format. For specific types of work packages that would call upon human resources with variable productivity, for example designing a manufacturing line quality control process, duration estimation requires experience and expertise.

The project manager and the team members request subject matter experts to use their experience to predict work package duration and costs.

This is commonly called the Delphi rounds—collection of estimates from a panel of recognized experts. Each expert is provided with a description of the work to be evaluated and the experts return their estimate.

The estimate does not need to be justified by the expert; however, it is helpful if it can be substantiated. The estimates from the first round are collated and distributed to the panel of experts. This process is repeated until consensus is reached. The project manager and the team members can then apply the mean estimate to the work package.

3.7.6.4 Vendor Bids

This technique is used for work that will be performed by an external organization—for example, subcontractors, suppliers, service providers.

The project manager and the team members provide a comprehensive description of the scope of work. The received bids, which will state principally the duration and cost, are collected and compared to ensure that all refer to the same scope of work. This is imperative in cases when the scope of work requests innovative and creative responses from the vendors.

Each organization will have its own procedure for vetting and screening vendor bids prior to contract award. Subsequent to the choice of vendor, the given duration and cost are posted for the work package(s) to be performed by the external organization. Other considerations for vendors will be presented in Chapter 6, Procurement Management.

3.7.6.5 Grass-Root

This technique, also called bottom-up, provides estimates with a high level of accuracy.

Bottom-up estimating requires a detailed WBS; however, it produces a more definitive and more accurate estimate.

The whole range of estimating techniques can be used depending on the nature of work packages—linear or variable productivity: analogy, parametric, subject matter experts, and vendor bids.

3.7.6.6 PERT Probabilistic

The program evaluation and review technique (PERT), presented as an activity network model, is a probabilistic estimation technique that allows for randomness in activity completion duration. A distinguishing feature of PERT is its ability to deal with uncertainty in activity completion duration.

This chapter will only address the probabilistic estimation technique of PERT. The networking facet is covered in Chapter 4, Project Scheduling.

PERT was devised in 1958 for the POLARIS missile program by the Program Evaluation Branch of the Special Projects office of the U.S. Navy, helped by the Lockheed Missile Systems division and the consultant firm of Booz-Allen & Hamilton.

PERT differs from the critical path method (CPM), in that CPM is a deterministic method that uses a fixed duration estimate for each activity. While CPM is easy to understand and use, it does not consider the duration variations that can have a great impact on the completion time of a complex project.

CPM was introduced by DuPont de Nemours & Co. and Remington Rand, circa 1957. The computation was designed for the UNIVAC-I computer. The first test was made in 1958, when CPM was applied to the construction of a new chemical plant.

For each activity, the PERT probabilistic model requires three duration/time estimates. Each estimate is qualified by the assumptions it considers (see Figure 3.27).

- Optimistic time (a):
    - The minimum possible time required to accomplish a task, assuming everything proceeds better than is normally expected
- Most likely time (m):
    - The best estimate of the time required to accomplish a task, assuming everything proceeds as normal
- Pessimistic time (b):
    - The maximum possible time required to accomplish a task, assuming everything goes wrong (but excluding major catastrophes)

The PERT estimation model assumes:

- The three given durations fit a beta distribution.
- The duration estimate is illustrated as a normal distribution.
- The span from optimistic to pessimistic covers six standard deviations.

PERT Probabilistic Model

Based Upon:
a - Optimistic
m - Most Likely
b - Pessimistic

Assumes a
Normal Distribution

FIGURE 3.27
PERT probabilistic model

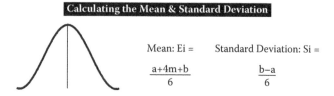

FIGURE 3.28
Calculating the mean and standard deviation

As PERT assumes a beta probability distribution for the time estimates, the expected time for each activity can be approximated as a weighted average:

(Optimistic + 4 × Most likely + Pessimistic)/6

And the standard deviation is approximated as:

(Pessimistic - Optimistic)/6

These deviations can be seen in Figure 3.28.

From the three estimates, and assuming a normal distribution, the probability of a duration range can be derived (see Figure 3.29).

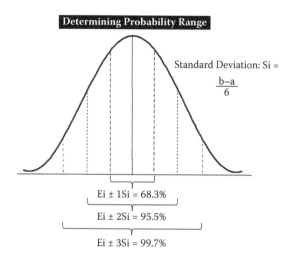

FIGURE 3.29
Determining probability range

Example

For the installation and testing of one CCTV, two experienced technicians may perform the work in an optimistic time of two hours if everything goes very well. From their previous experience, they usually will take 4 hours. If they encounter difficulties, such as problems with passing through cables or having to bypass structural elements, they may take up to 9 hours.

The mean time estimate can be calculated as:

$$(a + 4m + b)/6 = (2 + 16 + 9)/6 = 4\,h\,30\,min$$

The standard deviation = $(9 – 2)/6 = 1\,h10\,min$

When applying +/–1 standard deviation, there is a 68.3 percent probability of completing the work between 3 h 20 min and 5 h 40 min.

It is more common to state a "not to exceed" duration. By using the figures below:

$$Ei + 1\,Si = 84.15\%$$

$$Ei + 2\,Si = 97.75\%$$

$$Ei + 3\,Si = 99.85\%$$

Example

When applying +1 standard deviation, there is a 84.15 percent probability of not exceeding 5 h 40 min to complete the work.

PERT has a major weakness, in that the three time estimates are some-what subjective and depend on judgment. In cases where there is little experience in performing an activity, the numbers may be only a guess. In other cases, the person or group performing the activity estimates may introduce bias in the estimate.

3.7.7 Additional Estimation Allowances

The project manager and the team members must assess the estimates and make allowances for additional factors that cause cost variances.

3.7.7.1 Escalation

This considers the increase in cost caused by inflationary and market movements as measured from the cost estimate base date. This is the cut-off date at which wage rates, plant costs, and materials and equipment pricing utilized in the cost estimate development are based.

Cost estimates are either expressed as real terms or as money of the day (also called "nominal"). Real-term cost estimates are produced during the initial stage of the project development where forward escalation is not considered. Money-of-the-day cost estimates are produced during the planning and execution stages of the project and include forward escalation for the project duration. All cost estimates must be clearly labelled as either "real terms" or "money of the day."

Guidance on inflation and foreign exchange rates should be obtained from the financial division. Escalation for market movements tend to be localized and should be set in consultation with the appropriate departments.

3.7.7.2 Contingency

This is the reserve to be allocated for a cost estimate to allow for incomplete project definition, uncertain elements, omissions, and inadequacies that will bring the cost estimate up to the required accuracy.

Contingency is an integral part of the cost estimate, covering items and costs that cannot be defined at the time a cost estimate is produced. For all estimates, the level of contingency is assessed based upon the level of definition or detail available, market and historical data, contracting strategy, the apportionment of risk, and local knowledge. The level of contingency will reduce as the definition of the project improves.

The level of contingency associated with each element of the cost estimate should be clearly identified and documented.

Contingency is not intended to cover inconceivable disasters, such as major scope changes, unusual economic situations, extreme weather conditions, "force majeure," or strikes.

3.7.7.3 Indirect Costs to Consider

Each organization will allocate indirect costs to the project. These can be included in the hourly or daily rates of human or equipment resources as "fully loaded." Otherwise these are allocated by category, an example of which is presented in the table.

The project manager must ensure that the final project cost is coherent and includes whatever indirect costs are applied by the organization.

Category	Typical Scope of Services
Parent Company Overheads	Senior management/legal/commercial, includes cost of personnel (social security, national insurance, pensions, etc.)
Project Management Team	Personnel dedicated to the project including travel/relocation and subsistence costs
Project Support	Personnel assigned to support the project's administrative tasks
Office Support	Office rental/equipment and services
Communication/ Documentation/PR Services	Translation/project publicity materials/documentation archiving
Project Financing	Costs associated with third-party financing of the project in question
Land Acquisition	Land acquisition costs
Field Surveys & Support	Mapping and site investigation services and surveys
Certification	Certifying/warranty surveyors/vendor and fabrication inspection/governmental/ regulatory permits and approvals.
Insurance	Personal and consequential damages insurance premiums
HSE Support	Sustainable development /environmental /safety consultants /in-country support
Security	All costs associated with security—covering personnel/ camps, vehicles communication equipment and training
Information Technology and Communication	IT hardware/software/satellite/telephone communication systems

3.8 PRECEDENCE ANALYSIS

This section presents the precedence analysis concepts. Chapter 4, Project Scheduling explains network diagrams and the critical path method (CPM) in detail.

The project planning activity is concluded by determining the logical and physical dependencies of the work packages.

Only work packages are considered and ordered following a precedence analysis. The project manager and the team members will thus convert the hierarchal WBS to a sequenced network of activities (see Figure 3.30).

The network of activities provides the project team the flow of how the tasks are related, and conditions whether they can or need to be accomplished sequentially or overlap.

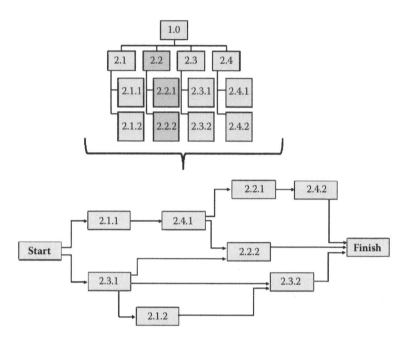

FIGURE 3.30
From WBS to activity network

A network is not cyclical—there are no loops, as every task must be connected to another task or event, thus creating paths over time.

The network is created based on assumptions about resource availability and estimated durations of each task, as well as on the evaluation of the interdependencies of work packages. During project scheduling and the construction of a calendar Gantt chart, some or all of these assumptions may have to be reassessed.

3.8.1 Determining the Sequence of Work Packages

The convention in building a network of activities is to have a "start" activity at the beginning and a "finish" activity at the end. Another convention is to show the flow of the network from left to right and not top to bottom.

Dependencies among work packages are defined and recorded for each. When predecessors are identified for each work package, they should only be for immediate predecessors. It is important to understand that multiple predecessors for a given work package are possible and often exist.

Other than the start and finish, all work packages will have a predecessor and successor work package. The network must also include all dependencies and relationships to work packages or events from and to external interrelated and concomitant projects.

The project manager must recognize:

- **Mandatory (physical) dependencies**—those dependencies that are inherent to the type of work being done. They cannot and will not change, no matter how many individuals are working on a task or how many hours are allocated to a task (e.g., on-site testing can only be performed after assembly and installation is complete). The project manager must recognize mandatory dependencies since they will dictate the way certain pieces of the schedule will need to be structured.
- **Discretionary (logical) dependencies**—those dependencies that are defined by the project team or customer, that offer the choice to the project manager to schedule tasks in a certain way. For example, the project team may decide to outfit the left wing of a building before the right wing, or vice versa, before installing any equipment.
- **External dependencies**—outside the realm of the project or outside the control of the project manager or customer, these dependencies may direct how portions of the project schedule must be defined. For example, a work package may be dependent upon another project's work package providing a piece of equipment or an approved document. This is something neither the project team nor the customer can control, but it must be defined and considered when revising the schedule.

3.8.1.1 The Activity-on-Arrow Network

The activity on arrow (AOA), which is used in both the critical path method (CPM) and program evaluation and research technique (PERT), is the historical basis for all activity network diagramming.

In the AOA network (see Figure 3.31), circles and arrows are used to represent the network logic. The circles are the events or nodes, and these are numbered sequentially. The arrows represent the work packages (activities or tasks) and they connect the nodes. The AOA is constructed respecting an established precedence analysis.

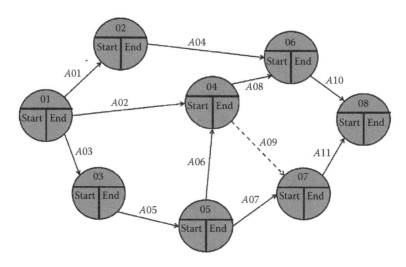

FIGURE 3.31
AOA network

The AOA can only use finish-to-start relationships. In addition, lead and lag times cannot be shown (this will be explained in Chapter 4, Project Scheduling).

The AOA requires *dummy* activities when the specific logical relationship between two particular activities on the arrow diagram cannot specifically be linked or conceptualized through simple use of arrows going from one activity to another.

As illustrated in Figure 3.31, because A11 depends on A06 and A07, but A08 depends only on A06, activity A09 is introduced as a *dummy* activity to preserve the logic for activity A06 (node 05 to 04) and A07 (node 05 to 07). Dummy activities are represented by a dashed line.

Finally, with the advent of project management software tools, and the wide use of the activity on node network, the AOA is now rarely used.

3.8.1.2 The Activity-on-Node Network

The activity on node (AON), often called the precedence diagramming method (PDM), represents activities as the nodes, usually in the shape of a rectangle, and the dependencies as arrows (see Figure 3.32). The resulting logic of the AON is the same as for the AOA.

The AON offers many advantages over the AOA: Dummies are not needed; dependencies are not limited to finish-to-start, as relationships

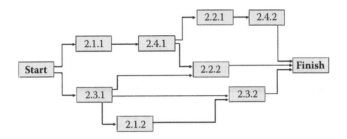

FIGURE 3.32
AON network

stipulating start-to-start; finish-to-finish and finish-to-start (rarely used) can also be introduced. In addition, leads and lags can be used.

Finally, all project management software tools support the AON technique.

3.9 CASE STUDY

3.9.1 Case Study Introduction

You are the manager of the marketing special projects group (SPG). This group offers marketing event management and support to the rest of the company. You report to the corporate SPG director, Jeremy DAVID.

Your company LEY-ZURE specializes in the manufacture and sale of leisure equipment. The company has over 2400 employees and yearly sales of ß5 billion (local currency is blicks).

Last week Will HEWITT, the marketing VP, requested the special projects team to assume the full responsibility for the organization and delivery of the company's marketing show at the next international trade fair. You have been assigned to manage this project—code name LEY-WORLD. The fair is a major occasion for your company to articulate its market strength and position and announce its latest products. This year the key focus will be on the LZB—the Portable/Kit Mountain Bike.

Market trials of the LZB have been very successful. Dave LOGAN and Justin REED, respectively R&D and manufacturing VPs, are extremely satisfied and are looking forward to the trade fair, where they will be present, as well as key members from their business units.

R&D is always seeking ways to ensure that requests from product marketing are in fact the real needs of the consumers. Some past products designed and developed have demonstrated that LEY-ZURE does not always respond clearly to market expectations. Dave LOGAN wants this new product to obtain a good market critique.

Manufacturing believes that it does not contribute as much as it could in the definition of new products or enhancements to current product lines. Justin REED often feels left out of strategic discussions between product marketing and R&D. His presence at the fair will help him to position his manufacturing view of things.

3.9.2 Case Study Scope of Work

Yesterday morning, Tuesday, November 5th, you attended a scope meeting for the event. Present at this meeting were Will HEWITT, Dave LOGAN, Justin REED, as well as the sales director Ernie KELLER, Sabrina MILET from product marketing, and Ivan DEEDS from the creativity group.

The following project scope was agreed:

Project LEY-WORLD

To Define, Design, Organize, Perform and Complete
the Company's "MARKETING PROMOTION WEEK"
at the next International Trade Fair in Trinidad and Tobago
starting May 15th NEXT YEAR

The major needs expressed can be summarized as follows:

1. Marketing wants to achieve a high impact on both current customer base and prospective new accounts. Product marketing wants to establish the key messages for the show.
2. Sales department wants to generate substantial leads, and establish potential business closures.
3. R&D wants to capture any expressed product enhancements.
4. Manufacturing wishes to seek out issues concerning quality, safety, and maintenance.
5. The creativity group requests sole responsibility in the design concept of the stand and booth.

Further to discussions concerning the overall logistics of the trade fair, the following major topics concerning specifically the LZB were defined:

1. Product marketing requests a video of a bike assembly, and a fully assembled bike on display on a rotating stand.
2. Sales requests a team present on the stand to perform a bike assembly and to have 100 bikes that can be sold at a special price after assembly. Creativity approves this and thinks it's great.
3. Manufacturing agrees on the sale of assembled bikes at the show, but selling price must cover at least the 50 ß unit cost per unassembled bike, which includes transport and assembly team at show.
4. R&D's suggestion to build a mini-track that buyers could use to try out the assembled bike was rejected by all.

Will HEWITT agrees to the above complementary LZB needs, but emphasizes that these requests are above the budget already assigned for project LEY-WORLD. They will only be acceptable if it can be demonstrated that they do not add supplementary costs to the company's budget for the show and can be autofinanced.

3.9.3 Case Study Major Deliverables

Yesterday afternoon, you were called in to a meeting with Will HEWITT. He stressed the importance of the success of this marketing event and gave you the key success criteria and results that had to be met for project LEY-WORLD.

Success Factors	Major Deliverables
Invitation of Top 100 Clients	Marketing package
Invitation of Top 25 Prospects	Travel package
Achieve > 20 New Business Leads	Equipment transfer package
Total Cost is within assigned marketing budget	Stand design and logistics package
Quality Stand acclaimed by press	International trade fair report

He reiterated that the additional subproject for the sale of 100 assembled bikes during the show would only get management approval if this did not generate extra costs. This additional subproject would be code-named Project LZB-Sales. Will HEWITT insisted that this subproject cover the costs it engenders.

You are requested to present the overall scope of the project LEY-WORLD and its sub-project LZB-Sales at the next management meeting is scheduled for the 13th of November. Subsequently the total project plan including costs, resources, and schedules are to be presented for management review on November 27th for final agreement on December 18th.

Will HEWITT has designated his assistant Lisa FERRIS to be your main interface for this project and answer any questions you may have.

You have access to the following organizational units and their representatives:

MARKETING VP	Will HEWITT
ASSISTANT to Marketing VP	Lisa FERRIS
PRODUCT MARKETING	Sabrina MILET
CREATIVITY GROUP	Ivan DEEDS
R & D VP	Dave LOGAN
MANUFACTURING VP	Justin REED
ASSEMBLY head	Lee BROWN
SALES DIRECTOR	Ernie KELLER

3.9.4 Case Study Key Information and Data

The budget for project LEY-WORLD has been identified. You can expect that the overall project plan you will present will be challenged to achieve the given constraint. However, the additional budget for the subproject LZB-Sales has to be calculated accurately to get management approval.

During the initiation and planning stages of the project you will work closely with product marketing, manufacturing, and creativity group representatives. These will bring experience and expertise for the planning of the project. You should also give special attention to the subproject LZB-Sales, and work closely with sales on the financial justification.

Estimated Lead Times (compared to similar trade fair past projects)

Sales, together with product marketing, needs to advise client guests no later than three months prior to the event

Product marketing requires four weeks for production of commercial documentation to be distributed at the event

Manufacturing requires three months to initiate and acquire transport and customs documentation for equipment & material dispatch

Manufacturing estimates two weeks for shipment of unassembled bikes to the event site

Manufacturing will make available staff at the event for bike assembly

Creative group requires two months for space reservation at trade fair, and two weeks for stand/booth installation

Creative group requires two months to create the demonstration video

Travel desk requires to pre-book air travel and hotel accommodation three months ahead of event, and one month prior to event for final reservations

3.9.4.1 Sub-Project LZB-Sales Specifics

In order to determine the revenue from the sale of assembled bikes at the event, sales has recently received a market study, which details the projected volume of assembled bike sales for a given unit price. Extracts from the market research are:

The research was conducted at the previous international trade fair with a panel of 100 people chosen from the 1500 visitors... The number of assembled mountain bike sales is directly linked to the sales price and the assembly quality.

Sale Price per Bike	Potential Sales
95 blicks	Maximum: 5 bikes
90 blicks	Maximum: 15 bikes
85 blicks	Maximum: 50 bikes
80 blicks	Maximum: 150 bikes

Manufacturing has informed you that unassembled bikes can be supplied by their business unit on internal transfer prices of 50 blicks per kit, including transportation to the final destination.

The internal cost per hour for each member of staff is 50 blicks. This is only applicable for all planning activities. All personnel at the event will be available at no charge.

4

Project Scheduling

4.1 CHAPTER OVERVIEW

This chapter covers the second major step in planning—the project scheduling.

The scheduling techniques for an activity network are presented in detail, with special emphasis on the critical path method (CPM) for determining critical path activities.

The construction of the project Gantt chart is explained, as well as the other supporting documents that together constitute the project plan and the approved project baseline.

4.2 SCHEDULING OVERVIEW

To effectively plan and control a project, the project manager needs to be able to process large amounts of data quickly and accurately to ascertain the complexity of the resulting schedule.

The activity network and the activity scheduled bar chart—Gantt chart—are two of the key scheduling steps, which are central to successful time management planning, cost management planning, and resource management planning.

From the activity network, a structure is created to allow for project control and tracking and monitoring project progress, earned value management, information processing, and project reporting. These topics are covered in Chapter 7—Project Implementation.

The activity network and the Gantt chart provide a highly structured and methodical approach to project scheduling. Along with the project

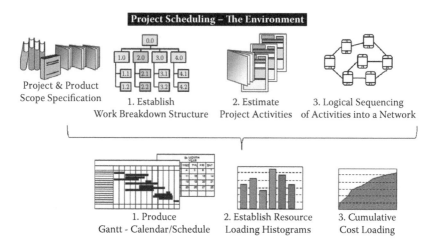

FIGURE 4.1
Project scheduling—the environment

resource loading charts and cost requirements, they are fundamental to establishing the project baseline plan (see Figure 4.1).

The scheduling steps continue from the project planning definitions given by the work breakdown structure (WBS) and the duration/cost estimations of work packages (for ease of vocabulary, scheduling will refer to activities— these are the work packages and/or tasks that have been defined).

The major steps are:

- Define the relationships (precedence analysis) between activities
- Draw the activity network diagram (with special attention to external interfaces to other project activity networks)
- Perform the network analysis (by determining absolute start and end of each activity and critical path)
- Transcribe network analysis results to a Gantt chart (obeying the civil and corporate calendars)
- Introduce project "must dates" (start and end of project, interfaced milestones to other projects, other external events)
- Optimize the activity network and resulting Gantt chart
- Build the cumulative cost estimate curve
- Establish and adapt the project's resource requirements

With the advent of project management software, there is a strong temptation to build the activity relationships and the network directly on the computer. The project manager must persuade the team members to avoid this, as not only is it extremely difficult to perform the precedence analysis on a computer screen, where few activities can actually be seen and the whole network be visualized, it most of all defeats all team dynamics objectives, as there is little or no active team member collaboration, participation, and eventual ownership. Use "brainware" and "teamware" first, and then use software.

Similarly, for the construction of the Gantt chart, project management software will open with a default screen, which "begs" for data entry of activities and their characteristics, such as duration estimation and connecting links. It is strongly suggested to avoid scheduling directly on the Gantt chart prior to developing a solid and logical activity network "offline."

4.2.1 Activity Networks

The first step before drawing the activity network is to determine the logical and physical dependencies of the work packages. Only activities (work packages) are considered and ordered following a precedence analysis (precedence analysis was introduced in Chapter 3). The project manager and the team members will thus convert the hierarchal WBS to a sequenced network of activities (see Figure 4.2).

A network is not cyclical—there are no loops, as every task must be connected to another task or event, thus creating paths over time.

The network is created based on assumptions about resource availability and estimated durations of each task, as well as on the evaluation of the interdependencies of work packages. As will be seen later in this chapter, some or all of these assumptions will have to be reassessed to map against the realities of resource availability during the construction and optimization of the calendar Gantt chart.

4.2.2 Determining the Sequence of Work Packages

The convention in building a network of activities is to have a "start" activity at the beginning and a "finish" activity at the end. Another convention is to show the flow of the network from left to right and not top to bottom.

Dependencies among work packages are defined and recorded for each. When predecessors are identified for each work package, they should only be for immediate predecessors. It is important to understand that multiple predecessors for a given work package are possible and often exist.

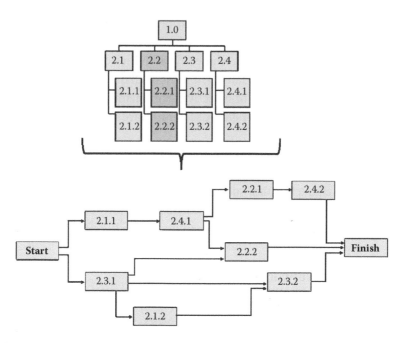

FIGURE 4.2
Transitioning the WBS to an activity network

Other than the "start" and "finish," all work packages will have a predecessor and successor work package. The network must also include all dependencies and relationships to work packages or events from and to external interrelated and concomitant projects.

The project manager must recognize:

- **Mandatory (physical) dependencies**—those dependencies that are inherent to the type of work being done. They cannot and will not change, no matter how many individuals are working on a task or how many hours are allocated to a task (e.g., on-site testing can only be performed after assembly and installation is complete). The project manager must recognize mandatory dependencies since they will dictate the way certain pieces of the schedule will need to be structured

- **Discretionary (logical) dependencies**—those dependencies that are defined by the project team or customer, that offer the choice to the project manager to schedule tasks in a certain way. For example, the project team may decide to outfit the left wing of a building before the right wing, or vice versa, before installing any equipment

- **External dependencies**—outside the realm of the project or outside the control of the project manager or customer, these dependencies may direct how portions of the project schedule must be defined. For example, a work package may be dependent upon another project's work package providing a piece of equipment or an approved document. This is something neither the project team nor the customer can control, but it must be defined and considered when revising the schedule.

4.2.3 Building the Precedence Chart of Activities

The project manager and the core team will build a precedence/relationship chart as follows:

Activity ID	Activity Description	Predecessor Activity ID	External Link: IN or OUT (identifying the activity, milestone event and/or deliverable)
PJ01-001	aaaaaaaaaaaaaaaa	START	
PJ01-002	bbbbbbbbbbbbbbbb	PJ01-007	
PJ01-003	cccccccccccccccc	X-xxxx	IN – Event: Complete infrastructure
PJ01-004	dddddddddddddddd	PJ01-001, PJ01-002	
PJ01-005			
PJ01-006			
PJ01-007			
PJ01-nnn	....		

The relationships from the above table can then be data entered into the project management software.

Often, and giving the most impact and teamwork, the project manager and the core team will construct and visualize the activity network on a wall, using Post-it notes either taken from the previous WBS structure or specifically developed for this step.

The activity network can then be illustrated either as an activity on arrow (AOA) or an activity on node (AON). This was explained in Chapter 3.

This chapter will concentrate on the activity on node (AON) as most, if not all, of the project management software packages use this presentation mode.

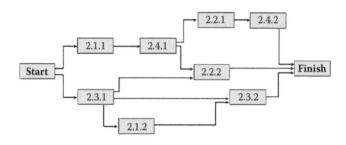

FIGURE 4.3
AON Network example

4.2.4. The Activity-on-Node Network

The activity on node (AON), also called the precedence diagramming method (PDM), represents activities as the nodes, usually in the shape of a rectangle, and the dependencies as arrows (see Figure 4.3).

This type of activity network offers a variety of relationships that are discussed in the next section.

4.2.5 Types of Activity Relationships

It is essential to establish the most effective physical and logical relationships for activities within the project's network, as well as to correctly define the interfaces from and to other projects. Only with a solid structure and framework can the activity network subsequently reflect the calendar for the project and its required resources.

The AON/PDM is very powerful, in that it offers four types of relationships that allow for more flexibility in defining the precedence between activities. Furthermore, the AON/PDM allows for meaningful ways to accelerate or delay the schedule, as will be seen in Section 4.2.7, Lags and Leads.

There are four relationships, sometimes called constraints, which are explained further in the following sections:

- Finish to Start—FS
- Start to Start—SS
- Finish to Finish—FF
- Start to Finish—SF

There is an additional relationship, a hammock, which is also explained below.

FIGURE 4.4
Finish-start relationship

4.2.6 Finish-to-Start Relationships

The finish to start (FS) is the most common relationship. An activity can only start if the preceding activity has been totally completed.

This relationship can be further developed by applying a delay between the two activities by the use of a lag (see Section 4.2.7.1).

In the following example, the walls of the room must be cleaned and completely dry before any painting can be performed. The FS link therefore shows that the "Paint Walls" activity cannot start before the "Clean/Dry Walls" is completely finished (see Figure 4.4).

4.2.6.1 Start-to-Start Relationships

The start to start (SS) represents the relationship between the start dates of two activities. Precedence still exists between the two activities, and the direction of the relationship is important, as the start of the successor activity is conditional only upon the start of the predecessor activity. During project implementation, any delay on the start of the predecessor will impact the start of the successor.

As for the FS relationship, this relationship can be further expanded by applying a delay between the two activities by the use of a lag. This is often used in fast-tracking a project (see Section 4.3.2.5).

In the following example, a total list of travellers and their final destinations has to be provided to the travel department, from which they can reserve the flight itineraries and trigger the accommodations reservation. The SS link shows that the "prepare hotel plan" activity can only start once the "prepare flight plan" has started (see Figure 4.5).

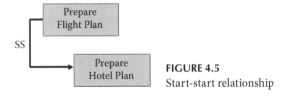

FIGURE 4.5
Start-start relationship

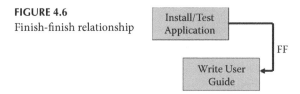

FIGURE 4.6
Finish-finish relationship

4.2.6.2 Finish-to-Finish Relationships

The finish to finish (FF) represents the relationship between the end dates of two activities. Precedence still exists between the two activities, and the direction of the relationship is important, as the end of the successor activity is conditional only upon the end of the predecessor activity. During project implementation, any delay on the end of the predecessor will impact the end of the successor.

As for the two previous relationships, this relationship between the two activities is often used with lag, and can also have a lead, and is very effective for fast-tracking a project (see Section 4.3.2.5).

In the following example, to ensure comprehensiveness, the user guide can only be completed when the entire application test has been concluded (see Figure 4.6). The FF link shows that the end of the "install/test application" activity is conditional on the end of the "write user guide."

4.2.6.3 The Use of Start-to-Finish Relationships

The start to finish (SF) represents the relationship between the start date of one activity and the end date of another activity (see Figure 4.7). Precedence still exists between the two activities, and the direction of the relationship is important, as the end of the successor activity is conditional only upon the start of the predecessor activity. During project implementation, any delay on the start of the predecessor will impact the end of the successor.

This relationship can be used when the start of the predecessor activity is the activating trigger to end the successor activity and/or when an

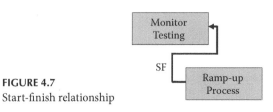

FIGURE 4.7
Start-finish relationship

external event constrains both the start and the end of the two activities. This constraint is often due to a "must date" imposed on the project (there is more information on "must dates" in the Section 4.3 on Gantt charts).

In the following example, the testing phase is performed for the duration of the phase and ends when it is judged that the system can be handed over to operations. The SF link shows that initiation of the "ramp-up process" activity will trigger the end of the "monitor testing" activity as project management monitoring and control is no longer required.

This relationship is seldom used.

4.2.6.4 Hammock Relationships

This is not a relationship *per se*. An extra activity, called a hammock activity, is introduced in the network to group a number of activities under one summary activity (see Figure 4.8). It is often used to insert a higher level key highlight for faster and easier reporting to senior management, who only wish to capture the project's important issues in the schedule at a summary level.

In the following example, the individual activities needed to complete the wall rendering are needed for detailed scheduling; however, management needs only to have a summarized view. A hammock activity "walls" is introduced (this can also be the upper "parent" level of these activities in the WBS). The "prepare walls" activity is linked to it by SS, whilst the "paint walls" has a FF relationship. Note the direction of the arrows that enter the "walls" hammock, indicating that the start and end dates are driven by the activities on the extremities of the path that it spans.

4.2.7 Lags and Leads

Lags and leads are techniques to introduce duration delays or accelerations in the network. These are used for all types of relationships. The logic

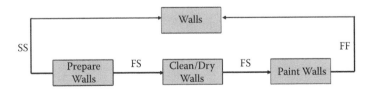

FIGURE 4.8
Hammock relationships

of predecessor/successor is respected; however, the start and end date calculations for each activity are adjusted according to the technique used (see below in Section 4.2.8, Forward and Backward Passes).

4.2.7.1 Lags

Lags will delay the successor's start or end date (depending on the type of relationship it has with its predecessor). Lags are often used when the successor activity, in order to commence, requires the predecessor activity to perform some of its own work (for example, once the core structural framework is constructed, the installation of utilities can commence). At other times, the successor activity can only be completed after receiving and utilizing the results of its predecessor. (For example, test results are needed and collated to draft the final project report.) Often, lags are introduced in the network to cater for a lapse of time that does not use resources (e.g., a six-week delivery time for a procured item).

The duration of the lag has to be included in the calculations of ES and EF.

4.2.7.1.1 Examples

Examine Figure 4.9, which shows examples of the application of lags.

- The FS relationship "clean walls" to "paint walls" includes a +5 lag to cover the wall drying time, which does not use any resources. This will delay the start of "paint walls" by 5 time units.
- The SS relationship indicates that 3 time units after the start of the "prepare flight plan" activity, enough has been done to initiate "prepare hotel plan."

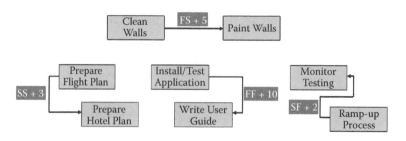

FIGURE 4.9
Examples of the application of lags

- The FF relationship specifies that 10 time units after the end of "install/test application" the activity "write user guide" can be completed as it can then incorporate results from the application tests.
- The SF relationship implies that 2 time units after "ramp-up process" has begun, the "monitor testing" activity can be completed to collate and finalize the test results.

4.2.7.2 Leads

Leads will accelerate the start date of the successor activity. They are only used for FS links when seeking to optimize and fast-track the project (see Figure 4.10).

In the following example, all the walls of a room have to be stripped before any preparation can begin. The logic constrains the "prepare walls" activity to follow the "strip walls," and therefore a FS relationship is needed. If, however, the work on "prepare walls" can begin once the majority of the work on the "strip walls" activity as been performed, then a lead can be applied. In this example the lead is of 3 time units. Thus the "prepare walls" activity can start earlier than it could if it had been a fixed FS link.

The duration of the lead has to be included in the calculations of ES and EF.

4.2.8 Forward and Backward Passes

The activity network represents the logical flow of work to be performed. A completed and verified network illustrates the sequences of activities. These sequences are called "paths." All the paths radiate from the start point of the network and converge back into the end point.

Now it is time for the activity network to return meaningful scheduling dates for each activity to the team. The critical path method (CPM) provides for this. The method was developed in the 1950s in a joint venture between DuPont Corporation and Remington Rand Corporation. While the

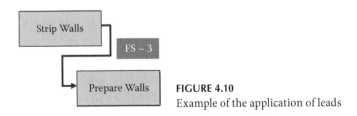

FIGURE 4.10
Example of the application of leads

FIGURE 4.11

Illustrating early and latest start-finish

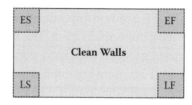

scheduling method was developed specifically for the construction industry, it can be applied to any project with interdependent activities.

CPM enables the project critical path to be determined by tracing the logical sequence of activities that directly affect the completion date of the project through a project network from start to finish. There may be more than one critical path depending on workflow logic. A delay to progress of any activity on the critical path(s) (without acceleration or re-sequencing) will cause the overall project duration to be extended.

Using CPM, date calculations are made for each activity along each path by proceeding by a "forward pass" of the network and then a "backward pass" (see Figure 4.11) (explained in Section 4.2.8.2). Each activity is then "boxed" by the following:

- **Earliest Start time—ES:** the earliest time at which the activity can start depending on its precedent activities.
- **Earliest Finish time—EF:** the earliest time at which the activity can finish. This is equal to the earliest start time for the activity plus the duration of the activity.
- **Latest Start time—LS:** the latest time at which the activity can start. This is equal to the latest finish time for the activity minus the duration of the activity.
- **Latest Finish time—LF:** the latest time at which the activity can be completed without delaying the project.

At this stage of scheduling it is important to note that all calculations will be made in absolute terms of duration in the established working unit, where the network's start is set to zero (0) (see Figure 4.12). For example, for a finish-to-start relationship, the absolute end date of the predecessor is at the same "date junction" as the start date of the successor (see Figure 4.13).

The Gantt chart, once adapted to both civil and corporate calendars, will present the schedule against a meaningful calendar.

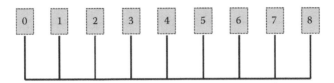

FIGURE 4.12
Absolute working unit framework

4.2.8.1 Forward Pass

The "forward pass" is the technique used to calculate the earliest start (ES) and earliest finish (EF) dates for each activity on each path of the network. It is a relatively simple process, requiring easy arithmetic (unless using PERT or other probabilistic estimating techniques).

The principle for single predecessor, finish-to-start (FS) relationships is straightforward:

- The network start activity, of a duration of zero, is set to zero. Its ES and EF are also set to zero. Proceeding down each path, the ES of an activity is set to equal the EF of the predecessor activity. The EF of the said activity is set to its ES plus its duration. This process is followed until all ES and EF dates of all activities are determined for all paths. As all paths converge to the finish activity, which has a duration of zero, the ES of the finish can be set and its EF set equal to the ES

4.2.8.1.1 Multiple Predecessors

When more than *one* predecessor exists, the ES of the activity will be set to the *highest* EF value of its predecessors. See the next paragraph for variations to this.

In the example in Figure 4.14, the "paint walls" activity cannot start before both activities "clean/dry walls" with an earliest finish (EF) date

FIGURE 4.13
Forward pass—determining early start and finish

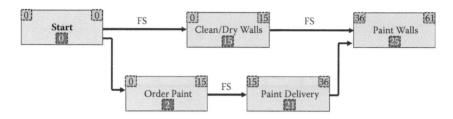

FIGURE 4.14
Forward pass—multiple predecessors

of 15 and "paint delivery" with an EF date of 36 are completed. By taking the highest of the EF dates, the resulting ES date of "paint walls" becomes 36.

4.2.8.1.2 Special Arithmetic for SS, FF, and SF Relationships

In *SS relationships,* the activity's ES will be set to its predecessor's ES. If a lag exists, then this is added to the successor's ES. The EF is then calculated by adding the activity duration to its ES.

In the example in Figure 4.15, it is estimated that the "prepare hotel plan" can commence 3 time units after the ES of the "prepare flight plan" activity. Thus its ES date is set to 20 + 3 = 23. With an estimated duration of 12, the EF date is set to 35.

In *FF relationships,* first the activity's EF is set to its predecessor's EF (see Figure 4.16). If a lag exists, then this is added to the successor's EF. The activity's ES is then determined by subtracting the activity duration from its EF.

In *SF relationships,* the activity's EF is first set to its predecessor's ES. If a lag exists, then this is added to the successor's EF. The activity's ES is then determined by subtracting the activity duration from its EF.

In the example in Figure 4.17, once the "ramp-up process" activity is initiated, the "monitor testing" activity can end following a wind-down of an estimated 2 time units. The EF date of the "monitor testing" activity is

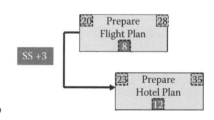

FIGURE 4.15
Forward pass—with start-start relationship

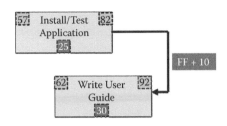

FIGURE 4.16

Forward pass—with finish-finish relationship

therefore set to 75 + 2 = 77, and for an estimated duration of 40, so its ES date is set to 37.

Please note that all the above examples are extracted illustrations, and that the calculations need to consider the effect of multiple predecessors and other types of relationships.

4.2.8.2 Backward Pass

The "backward pass" is the technique to calculate the latest start (LS) and latest finish (LF) dates for each activity on each path of the network. It is a relatively simple process, requiring easy arithmetic (unless using PERT or other probabilistic estimating techniques).

The principle for single predecessor, finish-to-start (FS) relationships is straightforward:

- The pass commences with the network's "finish" activity date. The ES date of the "finish" activity has been calculated by the forward pass. Since the activity has a duration of zero, then its EF date is set equal to the ES
- The LS and LF dates are also set to equal the activity's ES and EF [this is not always the case when "must dates" are set for the project's end delivery date—see Section 4.2.9, Calculating Float (Stack)]

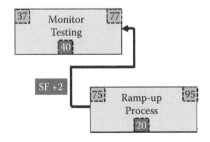

FIGURE 4.17

Forward pass—with start-finish relationship

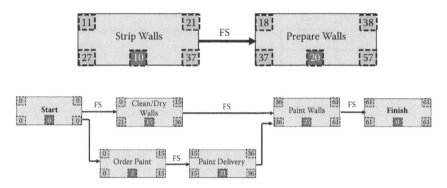

FIGURE 4.18
Backward pass—determining latest start and finish

- Proceeding backwards along each path, the LF of an activity is set to equal the LS of the successor activity. The LS of the activity is set to its LF minus its duration. This process is followed until all LS and LF dates of all activities are determined for all paths. As all paths converge back to the beginning of the network, the LF of the "start" activity can be set and the LS is set equal to it (see Figure 4.18).

4.2.8.2.1 Multiple Successors

When more than *one* successor exists, the LF of the activity will be set to the *lowest* LS of its successors. See the next paragraph for variations on this.

In the example in Figure 4.19, the "start" activity has two successor activities "clean/dry walls" with a LS date of 21 and "order paint" with a LS date of 0. By taking the lowest of the LS dates, the latest finish date of "start" is set to 0.

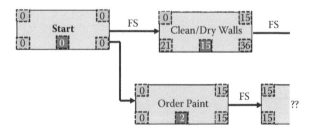

FIGURE 4.19
Backward pass—multiple successors

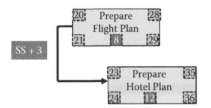

FIGURE 4.20
Backward pass—with start-start relationship

4.2.8.2.2 Special Arithmetic for SS and FF

In *SS relationships,* the activity's LS will be set to its successor's LS. If a lag exists, then this is subtracted from the successor's LS. The LF is then calculated by adding the activity duration to its LS.

In the example in Figure 4.20, working backward from the "prepare hotel plan" activity, the LS date of the "prepare flight plan" activity is set to 24 – 3 = 21. Then its LF date is set to 21 + 8 = 29, for the estimated duration of 9.

In *FF relationships*, the activity's LF is first set to its successor's LF. If a lag exists, then this is subtracted from the successor's LF (see Figure 4.21). The activity's LS is then determined by subtracting the activity duration from its LF.

Please note that all the above examples are extracted illustrations, and that the calculations need to consider the effect of multiple successors and other types of relationships.

4.2.9 Calculating Float (Slack)

Float is the amount of time that an activity can be delayed past its ES or EF without causing a delay to:

- Project completion date – TOTAL float (TF)
- Subsequent activities – FREE float (FF)

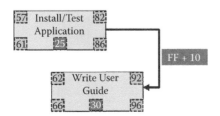

FIGURE 4.21
Backward pass—with finish-finish relationship

FIGURE 4.22
Determining total float

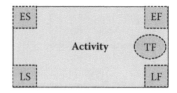

4.2.9.1 Total Float

The total float (TF) for an activity is the total amount of time that a schedule activity may be delayed from its EF date without delaying the project finish date, or violating a schedule constraint (see Figure 4.22 and Figure 4.23). Calculated as the time between its EF and LF time (the most common use), or between its ES and LS time:

$$TF = LF - EF \text{ or } LS - ES$$

4.2.9.2 Free Float

The free float (FF) is the amount of time that an activity can be delayed without delaying the ES of any immediate successor activity (see Figure 4.24 and Figure 4.25).

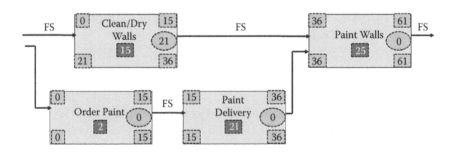

FIGURE 4.23
Total float—example

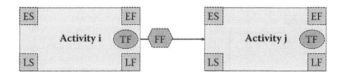

FIGURE 4.24
Determining free float

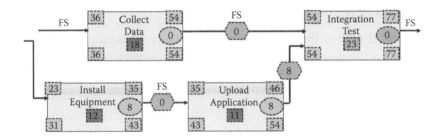

FIGURE 4.25
Free float—Example

The FF is determined between two consecutive activities and is the difference between the successor's (activity j) ES Start and the predecessor's (activity i) EF:

$$ES\,(j) - EF\,(i)$$

4.2.9.3 Implication of Float

A total float of zero indicates that the concerned activity is on the critical path (see Section 4.2.10). For non-critical path activities, the value of float, be it for TF or FF, gives the project manager the possibility of adapting the schedule to accommodate it to a variety of situations, such as delay on the start of the activity; unavailability of resources when required; longer duration than estimated, etc.

The FF has more significance than TF, since the former "absorbs" whatever TF is used. The limit of this absorption is the amount of FF. Any use above the amount of FF will affect the ES date of the successor activity. In Figure 4.25, whatever TF is used of the activity "install equipment" will affect the successor activity "upload application," as the FF of zero between the two cannot absorb any shift. The consequence would be that if any TF is used in the predecessor, it will have an incidence on the ES date of the successor activity.

Total or free float can be negative. This occurs when "must dates" are imposed as constraints. These constraints can be applied to the earliest start/finish and/or latest start/finish dates of activities. "Must dates" may be start or finish project deadlines, intermediate milestones that must be met, information available from a certain date, or other internal or external constraints imposed by interrelated projects. In essence, negative float means that the activity cannot realistically be performed in its current definition and estimate.

4.2.10 Determining the Critical Path

The critical path is the path through the project network in which each activity on the path has a TF of zero. This critical path is best illustrated when the activity network has been built as stand-alone and does not include any internal or external constraints. The network may have more than one critical path.

When constraints are applied to the activity network, TF may be negative, and this implies unrealistic dates that can only be met by modifying the definition and characteristics of the activities where possible.

The critical path is the longest-duration path through the network (as calculated by the forward pass, discussed above). The critical path conditions the shortest duration of the project—the project cannot finish before the calculated end date. Any delay on any activity on the critical path, a delay that cannot be absorbed by an explicit action or decision, will impact the project's end date.

In Figure 4.26, the critical path activities are "start," "order paint," "paint delivery," "paint walls," and "finish."

The project manager and the core team members must closely control the critical activities and take the possible and necessary measures to maintain the project's end date.

4.2.11 Optimization of the Activity Network

The project manager and the core team members should strive to optimize the activity network by applying, where possible and applicable, fast-tracking and crashing techniques (see Sections 4.3.2.4 and 4.3.2.5). Negative float, due to "must date" constraints, must be addressed by reviewing the durations of critical path activities and re-assessing the resource profiles and mix. It must be remembered that the activity network is still driven by an unlimited and unconstrained resource model.

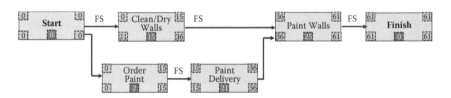

FIGURE 4.26
Determining the critical path

4.2.12 Building the Activity Network—Exercise

Below is a subset of the LEY-ZURE project and concerns the fabrication of the marketing support materials and the stand to be used at the trade fair. From the list of activities, draw the activity network and determine the critical path activities after performing the forward and backward passes and calculating the total and free floats.

Act Id.	Activity Description	Duration in days	Predecessor
01	START	0	
02	Project start-up planning meeting	3	01
03	Develop corporate brochures	6	02
04	Procure corporate binders	2	02
05	Design product flyers	5	02
06	Design stand	4	02
07	Produce corporate brochures	9	03
08	Assemble corporate binders/brochures	20	04,07
09	Produce product flyers	6	05
10	Procure outfitting equipment	11	06
11	Build stand platform	19	06
12	Assemble marketing materials	1	08,09
13	Assemble outfitting equipment	3	10
14	Lay utilities in stand	9	11
15	Mount stand partition walls	12	14
16	Test out-fitting equipment	7	13
17	Pack marketing materials	4	12
18	Finalize fittings	15	15
19	Integrate and ship stand	13	16,18
20	Ship marketing materials	10	17
21	FINISH	0	19, 20

4.3 BUILDING THE GANTT CHART

The Gantt chart is named after Henry Laurence Gantt, A.B., M.E. (1861–1919), an American mechanical engineer and management consultant who, with Frederick W. Taylor, applied scientific management principles to their work at steel plants (for more information, please search using your favorite Web browser). Gantt actually built charts to manage

ID	⊖	Task Name	Duration	January					February			
				28/12	04/01	11/01	18/01	25/01	01/02	08/02	15/02	22/02
1		**Start**	0 Days									
2		Define Test Requirements	6 Days									
3		Establish Test Location	3 Days									
4		Prepare Test Location	8 Days									
5		Define Test Controls	7 Days									
6		Write Test Plans	5 Days									
7		Prepare Test Bed	10 Days									
8		Equipment Installation	8 Days									
9		Integration Testing	6 Days									
10		Acceptance Testing	8 Days									
11		**Finish**	0 Days									

FIGURE 4.27
The Gantt chart

production schedules, with horizontal bars to illustrate durations of industrial process activities.

A Gantt chart illustrates a project schedule against a calendar by representing each activity as a bar stretching horizontally from its start date to its end date (see Figure 4.27). It answers the question "When in time will an activity be performed?" The activities are the work packages used in building the activity network, plus any hammock activities created during that process. Gantt charts can also illustrate the summary elements of a project as described in the work breakdown structure of the project. Milestone events can also be shown on these charts.

The Gantt chart has to be built in alignment to, and in consideration of, the civil and corporate calendars and the corresponding legal work-day hours, so as to have a valid realistic value. It will show the activities' earliest start and finish times (or conversely the latest start and finish times) in real dates, and no longer in absolute terms as was the case during the activity network step.

Gantt charts are the *outputs* of the planning and scheduling phase. They are the most-used graphic representations in projects. A common error is to build the Gantt chart by attempting to define the project work breakdown structure at the same time as defining the schedule and relationship of activities. This practice makes it very difficult to follow the 100 percent rule. The Gantt chart can only be produced after the scope of work, work breakdown structure, activity estimation, and activity network have been built and verified—these are the *inputs*.

Obviously, this is the appropriate time to use project management software, as the amount of available project planning data will be cumbersome to manage manually.

The project manager and the core team members should not be surprised to discover that the activity durations on the Gantt chart are now represented in "elapsed time," where civil calendars, weekends, public holidays, "must date" constraints, and other stoppages have been incorporated. It is at this time that the team can appreciate the difference between effort, work hours/days, and the resulting calendar elapsed time.

The Gantt chart is to be assessed and reviewed continually by the project manager and the core team members to optimize the chart and resolve key project "must dates" conflicts, lack of resource availability, and other conflicting constraints. This will also be done throughout the project execution phase as project schedule progress is recorded and actions taken to address variances to the project delivery.

It must be noted that Gantt charts only represent one of the project's constraints—schedule. The charts do not represent the size of a project, the resources required, the funding, or any of the risk actions that need to be undertaken. Plotting performance progress during project execution may be misinterpreted, as the activities on Gantt charts tend to be read as meaning constant, continuous, and linear productivity. This is certainly not the case when the majority of the work is knowledge-driven.

4.3.1 Establishing the Project Calendar

The Gantt chart must make realistic sense. The project calendar must be established in a comprehensive manner so as to reflect the real start and finish dates of activities and events. It is strongly suggested that project management software be used to capture the following elements that drive the project calendar:

- Details of the civil calendar to use, and how the legal work-week is distributed. Note that there are three major five-day working weeks: Monday to Friday; Sunday to Thursday, and Saturday to Wednesday. The civil calendar must also include public and religious holidays.
- Details of the corporate calendar to use, and the working-day legal hours scheme of human resources by type of profile. The corporate calendar must also include established company stoppages, such as office and factory closures, and cater for known low-productivity periods, such as vacation months.

The calendar is then adapted to include:

- Principal deadline dates such as start of project and end of project
- "Must dates" for project key deliverables and milestones
- Fixed dates for interfaces to external projects

The realistic and date-constrained project calendar is now ready to receive the scheduling information developed by the activity network.

4.3.2 Meeting the Schedule Constraints

The result of the activity network is now mapped into the project calendar. Primary key steps are to establish the assumed or given project start date, and to introduce the given and known project end date constraint. Thus, the project calendar is bound by a time window in which the project must be performed.

The usual practice is to use the earliest start and finish dates when positioning the activities on the Gantt chart. This will place whatever total float exists to follow the activity. Some may prefer to lay out the activities using the latest start and finish dates, and have the total float precede the activity.

A project's schedule cannot be considered as a stand-alone object; it has to relate to company objectives, strategies, and priorities. Combining all the schedules of the different projects is a matter of acute complexity for managers at all levels in the resolution of resource allocation and optimization while meeting group objectives.

Scheduling rounds convert an initial schedule (based upon estimations computed on infinite resources) into an approved baseline project schedule, which is constrained to finite resource availability and calendar. These rounds will invariably have to resolve the conflicts arising from the completion of all project objectives within limited funding and imposed delivery dates.

Two techniques are used in order to resolve scheduling constraints—activity-driven and resource-driven. It must be remembered, however, that the project must establish a policy defining which of the two approaches is to be used. They are by nature in conflict with each other, so it is rare for a project to be both activity and resource driven and achieve the same time schedule.

Therefore, the project manager must adopt the approach that is most suitable to meet the scope, financial, and schedule performance constraints.

This is of critical importance so that conflicts and trade-offs resulting from these constraints can be resolved as effectively as possible.

The network constructed from the project's activities is logical. This means that once built, it imposes the sequence of work. Therefore any shift applied to an activity will result in shifting its successors.

4.3.2.1 Activity-Driven Schedule

The first Gantt chart produced is activity driven. The schedule is based on the activity necessities under an unlimited and constrained resources model.

Activity-Driven Schedules

They drive the schedule when sufficient resources are assigned to each activity.

In the activity-driven schedule approach, the activities drive the schedule, and resources are to be made available so that project objectives can be reached within set time constraints.

The project manager and the core team members will analyze the resulting activity-driven schedule and address the incoherence found for activities with negative total float. This implies that the late finish date of the activity is before the early finish date and is the result of the imposed date constraints, such as deadlines and "must dates." It is obvious that this is an unworkable situation that the team will need to resolve by applying crashing and fast-tracking techniques where possible, as explained in Sections 4.3.2.4 and 4.3.2.5. Multiple iterations may be needed to arrive at a satisfactory end result.

The project manager will need to accept that in many cases an imposed project end date cannot be met even with unlimited resources.

4.3.2.2 Resource-Driven Schedule

Having established an optimized activity-driven Gantt chart, the project manager and the core team members will proceed to secure the required resources by profile. The project will be confronted by the reality, and the future availability, of resources. Limited and constrained resources will alter the Gantt chart to be resource driven—the schedule is based on the availability of resources, and the activities will need to be shifted in time to accommodate these limitations.

Resource-Driven Schedules

These are driven by resource availability and may cause activity schedule slippage to accommodate any imposed constraints.

Using project management software, the team will have produced resource loading charts for each resource profile, as identified during the estimating rounds. Each histogram represents the volume of resources required by corresponding time units to the Gantt chart (see Figure 4.28).

The challenge for the project manager is to establish, with all internal and external resource providers, the engagement and commitment to make available the requested resources, according to the needed profiles, for the period indicated for the activity on the Gantt chart. When resources are unavailable, or not of the required profile, then the activity has either to be shifted to the future or has to be reassessed for its contents. Reestimating may need to take place, which may result in an extended activity duration and corresponding increase in cost.

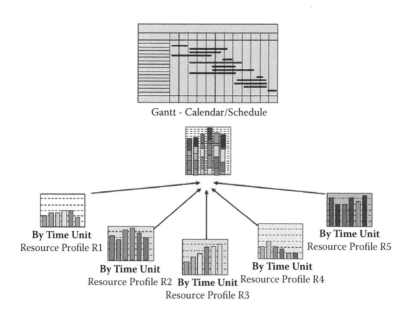

Gantt - Calendar/Schedule

By Time Unit
Resource Profile R1

By Time Unit
Resource Profile R5

By Time Unit
Resource Profile R2

By Time Unit
Resource Profile R4

By Time Unit
Resource Profile R3

FIGURE 4.28
Synchronizing resource requirements

The consequences of shifting activities, or of longer duration, is severely felt for activities on the critical path, as the project end date will be shifted and "must dates" on the critical path may produce a negative total float.

Noncritical path activities can only be shifted according to the available free float without impacting on the start of the successor activity and eventually their convergence with the critical path.

Multiple iterations will be needed to arrive at a satisfactory end result. The project manager may well need to seek assistance and support from the project sponsor and key stakeholders to modify resource priorities, so as to satisfy the project's deadline.

The project manager will need to accept that in many cases an imposed project end date cannot be met with limited resources.

4.3.2.3 Trade-Off Analysis

In some instances, the resource-driven schedule may be feasible, with secured resources committed, and all the deadlines and "must dates" able to be met. The project manager can then proceed to project plan and baseline approval. This is rarely the case.

As explained previously, both activity-driven and resource-driven schedules will test the imposed constraints to their limit, and often will not abide by the project's deadlines, as no project exists in a vacuum.

"Leveling" is attempted in order to optimize the available floats. If this is not successful then "constraining" will force activity slippage resulting in delays to the project completion date. If intermediate "must dates" are part of the initial schedule, interfaces to other projects and activity slippage will show that they will not be reached. It will then become apparent that a major conflict exists and that the schedule can only be completed following trade-off negotiations.

The project manager must try to resolve these conflicts by establishing multiple proposed schedules. These will be used later with management and partners who will select a suitable schedule.

The project manager must now enter a trade-off analysis with the sponsor and key stakeholders and discuss which of the three major vectors of the triple constraint needs to be adapted to the reality of the situation.

- Can the scope of work be reduced, thus decreasing the work load and fulfilling both the cost and duration constraints?

- Can the project deadline be extended to accommodate the lack of available resources?
- Can the project cost be increased to fund other ways of securing resources, which may be more costly?

Eventually, can the project be suspended or even cancelled when any of the three major options is not acceptable?

Once the negotiated schedule and resource histograms are finalized, they are submitted for approvals, resource contracts, and commitments from the corresponding resource managers. When they are signed off, the baseline project schedule can be published.

4.3.2.4 Crashing

Crashing is a schedule compression technique that reduces the duration of an activity. For maximum impact on the overall schedule, it is best applied to the activities on the critical path.

Crashing an activity would consist of:

- Allocating more resources
- Modifying the type/quality of the resource mix
- Extending normal legal work hours
- Shortening delivery from providers by renegotiating terms of payment
- Reducing the scope of the activity
- Breaking down the activity to create overlap of subactivities

The project manager must conduct a trade-off analysis to ascertain the schedule gain compared to the associated increased costs incurred.

4.3.2.5 Fast-Tracking

Fast-tracking is a schedule compression technique to shorten the duration of a path of activities. To have maximum impact on the overall schedule, it is best applied to the activities on the critical path. Fast-tracking consists of reviewing the FS relationships and determining ways to overlap the activities. The precedence logic between the activities is maintained.

The most usual ways to introduce fast-tracking are:

- Establish a start-to-start (SS) relationship rather than its original FS. A lag can also be added where necessary (see Figure 4.29).

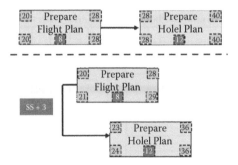

FIGURE 4.29
Fast tracking example 1

- Establish a finish-to-finish (FF) relationship rather than its original FS. A lag can also be added where necessary (see Figure 4.30).

4.3.6 Drafting the Milestone Schedule

Project scheduling considers a vast amount of information and details. This is necessary for all project team members; however, it is too cumbersome for sponsors, key stakeholders, and management in general.

Summarized Gantt charts are produced by the project manager, according to the audience targeted, which includes project managers of interfaced projects and external providers. The milestone schedule is the most used (see Figure 4.31). This depicts when in time a certain key event of the project will take place—for example "start of hand-over."

At times the milestone schedule may be too high level and/or cryptic. The summary Gantt chart can then also represent the major activities or phases that precede the milestone, and the major project deliverables.

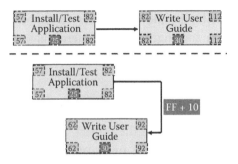

FIGURE 4.30
Fast tracking example 2

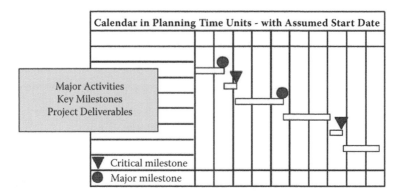

FIGURE 4.31
Milestone schedule

4.3.7 Building the Gantt Chart—Exercise

From the activity network developed for the LEY-ZURE project (see Section 4.2.12, Building the Activity Network—Exercise), and the calculated start and end dates, establish a Gantt chart, positioning the activities to their earliest start and finish dates.

Assume the project start date is January 4th, 2010; the normal work week is five-day, Monday to Friday with an eight-hour legal day; there are no public holidays or any other stoppage days for the duration of the project.

What is the project calendar finish date?

4.4 BUILDING THE COST ESTIMATE

The funding of the project is governed by the total costs estimated for the work to be performed, plus additional reserves to cover risk contingency plans and estimation contingency, and general management reserves to address the unknowns.

4.4.1 Establishing the Project Cost and Funding

The project manager establishes the project cost breakdown by initially aggregating the direct costs. These are driven by the cost of resources and services/fees related to the work activities. Indirect costs must also be included. Each corporation would have a different technique to distribute

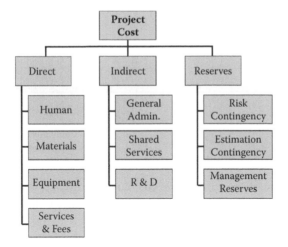

FIGURE 4.32
Cost distribution

its overheads. Care should be taken by the project manager not to double-count costs, especially when the internal hourly-rate direct cost has already been "loaded" with an overhead (see Figure 4.32).

The reserves have to be negotiated and agreed with the sponsor and key stakeholders. Risk contingency is established from the risk analysis and response step (see Chapter 5, Risk Management). The estimation contingency is agreed to cover the estimation error range, while the management reserve is usually a lump sum made available to cover unknowns. The project manager must seek to secure agreement on reserves so that these can be called upon when the need arises. Without previous agreement with the sponsor and key stakeholders, the project manager would need to negotiate a funding extension during project implementation, which may be a lengthy process, especially at a time when speed is required.

4.4.2 The Cumulate Cost Curve

The project manager establishes a cumulative cost curve that represents the financial "burn rate" requirements of the project (see Figure 4.33).

The curve will serve primarily to secure funding for the total project. In many cases, the project schedule can be divided into convenient periods of time (monthly or quarterly) or when key deliverables and milestones are set. At these milestones, management can review project progress and release funds for the next period.

FIGURE 4.33
Cumulative cost curve

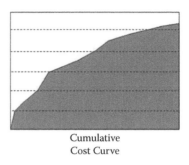

Cumulative
Cost Curve

The major use of the cumulative cost curve, however, will be to monitor and evaluate project progress during the implementation phase. This is discussed in Chapter 7, Project Implementation.

4.5 RESOURCE MANAGEMENT

The project schedule cannot be concluded, nor approved, until such time as the appropriate resources are allocated to the project. The major challenge is to secure the various types of resources from a variety of sources—the resource providers: internal from the performing organization, and external from contractors, suppliers, service providers, and third parties.

The project manager must align the project resource needs to the reality of resource availability, and negotiate the assignment of resources by reaching agreement and commitment with the resource providers.

The resource providers are:

- Functional managers responsible for human resources who can contribute to the project and have the required skills and competence profile. These managers must also consider the net availability of human resources who may have involvement in operational work.
- Line managers responsible for the allocation and distribution of asset resources such as equipment.
- Providers of services.
- Suppliers of products.
- Contractors for turnkey projects.

The project manager must discuss the resource requests with all the resource provider entities and adapt the project schedule accordingly.

The backdrop for resource assignment negotiation is the priority that has been given to the project. The project manager must seek the assistance of the sponsor and key stakeholders to secure resources where necessary else agree and approve changes to the major project constraints, namely scope, cost and/or schedule.

4.5.1 Resource Planning Techniques

These techniques have been explained in this and previous modules, and principally consist of clearly identifying, during the estimation rounds, the resource profiles and their characteristics. The volume of required resources by profile is graphically represented—this is best done by a histogram, but can also be done in a tabular format.

The iterative process between activity-driven and resource-driven schedules requires project management software to represent quickly the consequences of resource constraints.

4.5.1.1 Resource-Loading Table and Histogram

Each resource profile is represented with a resource-loading histogram, substantiated by the project-planning documentation (see Figure 4.34).

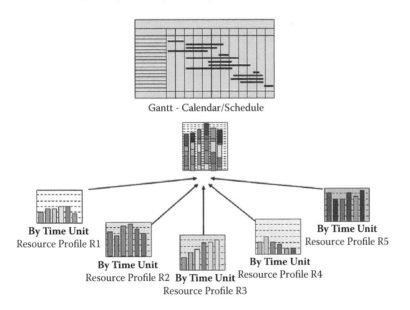

FIGURE 4.34
Resource loading histograms

All histograms are used to negotiate and secure resource assignment. Due to resource constraints, it is imperative, following each iteration, to reproduce the histograms and measure the impacts to schedule and cost.

4.5.1.2 Resource Leveling

Resource leveling is an optimization process used by the project manager and the core team members to examine a project for an unbalanced use of resources over time, and for resolving overallocations or conflicts. Leveling is first used to smooth the peaks of the resource histograms to maintain constancy of resources as much as possible—the project manager cannot have a situation in which resources are allocated in bursts and spurts. Resource leveling is subsequently used to constrain or "cap" the resource histogram with the availability.

The consequence of resource leveling is the shift of activities further down in time and the constitution of a new resource histogram for all profiles affected.

Leveling must first be applied to activities that have total float until they exhaust whatever free float exists between consecutive activities on the same path. This action will avoid impacting on the critical path and can thus preserve the target project end date. If the first iteration of leveling does not return an acceptable result, then critical path activities are considered, with the obvious impact to the project end date.

Resource leveling is often a wake-up call to many managers as to the issues relating to limited and constrained resources and the inability to fulfill the triple constraint without foregoing one of the project constraints: scope, cost, or schedule.

4.5.2 Resource Allocation

Negotiating rounds with resource providers must be concluded with firm commitments to make the resources available to the project at the agreed described time. The agreement is not to allocate a named human resource or certain specific equipment, but rather a commitment that the resource manager engages in providing the agreed resource profile to the quantity requested.

4.5.3 Resource Provider Commitments

The project manager must seek signed internal agreements, or the equivalent, that will protect the resource allocation of the project from resource providers. Without these formal agreements, obtaining project plan approval from management will be meaningless, as there would be little or no guarantee of project delivery success.

The project manager would draft delivery contracts with external providers, which will be legally binding upon both parties (further explained in Chapter 6, Procurement Management).

Agreement must also be reached between the project manager and the resource manager on the process to follow for the subsequent release of resources during the project implementation. This agreement would consist primarily of the means to follow up on the status of both the project progress and the state of availability of the resources prior to their assignment to perform the corresponding activity.

4.5.4 Roles and Responsibilities—RACI Chart

To conclude the agreement and assignment of resources, the project manager will draft a matrix that highlights the roles and responsibilities of the functions assigned to perform on the project. This matrix can be a mirror of the WBS or can represent the project deliverables. The first draft of the matrix does not contain any names of human resources or any other nominative or specific identity for other resources. Once the actual resource to be assigned to perform on the project is known, the matrix can be adapted to insert the nominative information.

The RACI chart (see Figure 4.35) is commonly used to illustrate the roles and responsibilities matrix:

R—the resource **responsible** to perform the work
A—the individual **accountable** for the work to be performed
C—the individuals who are **consulted** to participate in the performance of the work
I—the individuals who will be **informed** about the work to be performed

4.6 PROJECT COMMUNICATION PLAN

Frequent and comprehensive communication is a key project success factor. The project manager is the pivotal point of communication to the project team, the sponsor, and all stakeholders, namely: internal functional

Matrix of Responsibilities

W.B.S. Group Element	Functional Resp. Functional Areas						
		1	2	3	4	5	6
1.0	Product Definition				X		
2.0	Design		X				
3.0	Layout				X		
4.0	User Training					X	
5.0	Acceptance	X	X			X	X

Where X = R, A, C or I

FIGURE 4.35
RACI chart

managers who will provide resources, other internal departments, and all contributing external organizations. (Stakeholder identification and analysis was explained in Chapter 2.)

The project manager will use the stakeholder communication plan to engage in networking with influential stakeholders, to gain and sustain support from advocates, and to achieve buy-in from critics (see Figure 4.36). The project manager must strive to communicate in stakeholders' "language."

The communication plan will cover the information flow from and to the project. The plan is comprised principally of project performance reports: scope, costs, and schedule and project status reports with progress and forecasts. Additionally, the communication plan

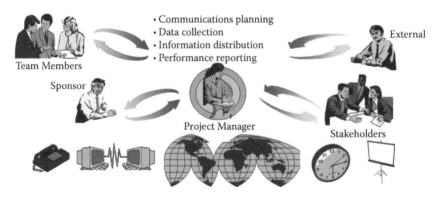

FIGURE 4.36
Project communication planning

will include project presentations, meeting minutes, change request status, issues, and other pertinent information that needs to be circulated within the project.

The communication plan must illustrate how the information needs of all project team members and stakeholders will be satisfied and verified with a feedback loop. The plan defines for each stakeholder the type of communication most likely to be effective, specifying the timing and frequency, the medium and the expected outcomes.

The results of the stakeholder communication plan are integrated into the project communication plan and the project schedule. During project implementation, they are monitored through team meetings and regular reports.

4.6.1 Establishing the Stakeholder Communication Map

As was explained in Chapter 2, stakeholders are identified and classified into categories. The type and content of communication will differ for each category (see Figure 4.37).

4.6.2 Determining Information Contents

The project manager will identify the whole set of information to distribute and establish a list of documents to produce throughout the project life cycle.

FIGURE 4.37
Stakeholder communication map

The list below is not exhaustive and should be modified and adapted to meet the specific project characteristics:

Communication Plan	Project Meetings Reports
Statement of Work	Project Performance Report
	Project Progress Report
Project Organization Chart	Project Financial Report
Project Plan	Earned Value Performance Report
Work Breakdown Structure	Project Closure Report
Project Cost Estimates	Project Quality Audit
Project Updated Schedule	Inspection Reports
Roles and Responsibilities Matrix	Change Requests Report
Resource Requirements	
Resource Training Plan	Risk Management Plan
	Risk Assessment
Decision Requests	Risk Response Plan
Steering Committee Minutes	
	Project Risk Report
Contracts and Terms and Conditions	Project Issues

4.6.3 Selecting the Appropriate Communication Media

The project manager should chart the most appropriate communication media to use depending on the nature of the project and the target audience. As can be seen in Figure 4.38, the media also carries the goal of the communication, from awareness through to commitment. The chart depicts this.

4.6.4 Drafting the Communication Plan

The communication plan is then constructed and contains the distribution list of project documents, their destination, and attributes, such as medium and frequency (see Figure 4.39). This plan is to be discussed and agreed with the involved individuals who are destined to receive the documents.

4.6.5 Determining the Types of Meetings and Their Agendas

As part of the project's communication plan, face-to-face meetings will consume at least 50 percent of the project manager's time. A meeting

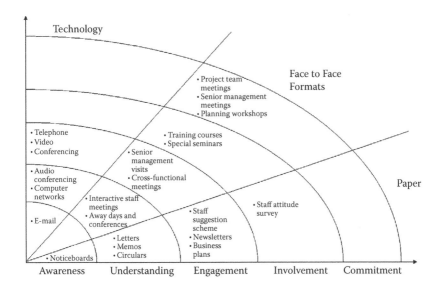

FIGURE 4.38
Project communication media

plan must be established and posted in the project calendars of all participants.

There are various types of project meetings depending on their nature, objectives, or purpose. It is strongly suggested that meetings are clearly named and cover only the topic described. Meetings combining multiple purposes, requiring a disparate audience, and an extended time frame are unproductive and waste most of the audience's time. If the project manager has to hold many meetings in a short space of time, for example when present at an outside location, then it is recommended that multiple

Stakeholder	Objectives	Type of Communication	Medium	Frequency and Schedule	Distribution	Outcome
xxxx	xxxx	xxxxxxxx	xxxx	xxxx	xxxx	xxxx

FIGURE 4.39
Project Communication Plan

one-topic meetings be held in a series, when different audiences can be invited and be more focused, participative, and active.

4.6.6 Meeting Types

The table in this section presents a nonexhaustive list of project meetings by purpose. The project manager can plan most of these in advance, and post them in the project calendar. Other meetings will arise during the course of project implementation. Each meeting needs to target the appropriate audience, for example internal with the team; sponsor and key stakeholders; functional managers and external with client; providers and suppliers.

Type of Meeting	Purpose
Ad-hoc	Team members gather together for a special purpose
Brainstorming	To obtain a comprehensive list of ideas and thoughts on a particular issue or topic
Consultative	Held with one or more experts to understand a particular issue and comment or advice on it
Decision-making	Conducted to make an agreed decision about an issue
Electronic	Virtual meetings such as conference calls, Internet-based voice or video conferencing
Informal	Held in a casual manner and atmosphere around a coffee machine, at lunch, or dinner
Kick-off	Initial project or phase start-up
One-on-one	Discussion between the project manager and another individual
Planning	Led by the project manager to plan or re-plan the project
Problem-solving	Dedicated to addressing specific problems
Status and Progress	Reporting to management on status and progress
Team	Repetitive events held at predefined times such as weekly, monthly, etc. These often cover individual status, project status, team status, discussion of common problems, etc.

4.6.7 Meeting Planning Framework

Formulating a set of objectives in the meeting preparation phase is the first and most important step to keep the participants focused on what they need to accomplish in the session. The objectives have to be realistic and measurable to become achievable. Effective meetings require preparation, participation, and follow-up.

Meeting preparation actions include:

- Determine the objective or purpose of the meeting
- Select the participants ensuring that the group is limited to a small number to facilitate discussion
- Set the agenda and the timing for each agenda item to ensure that the meeting can be completed within time constraints
- Distribute the agenda, and background information, to participants ahead of time to provide adequate time for preparation.

Meeting participation with clearly defined roles includes:

- Project manager to guide the discussion and include all participants
- Project manager to ensure the meeting starts and finishes on time
- Minute taker to record decisions and action items and prepare meeting minutes.

Meeting follow-up includes:

- Distribute meeting minutes to participants and others concerned by the discussion items
- Follow-up on progress on action items
- Schedule additional meetings as required

4.6.8 Meeting Agendas

Each type of meeting should have a standard agenda template that can be modified to accommodate specific topics.

A typical project status and progress meeting agenda is shown here.

Welcome and presentations
1. Nominate secretary
2. Restate meeting objectives and agenda
3. Agree on points of discussion (*issued by letter*)
4. Identify additional issues
5. Establish timeframe/priorities of meeting
6. Review outstanding issues from previous meeting

7. Review project progress
    schedule performance
    milestones status
    activity status
    scope performance status
    financial performance status
8. Establish next period milestones
9. Establish list of changes:
    incorporated
    to investigate
10. Document issues and actions (*distribute if possible*)
11. Summarize decisions (*distribute if possible*)
12. Conclusion and review of initial minutes (*distribute if possible*)
13. Agree on next meeting
14. Farewells

4.7 CONSOLIDATING THE PROJECT PLAN

4.7.1 Assembling the Project Plan

Prior to proceeding to seek management agreement and project baseline approval, the project manager will collate all the components, which constitute the project plan.

A sample table of contents is included here:

Management summary
   Executive summary
      Business need
      Project charter
      PM designation letter
      Project key stakeholders
      Proposed high-level solution
      Gantt key milestone schedule
      Funding requirements
      Acceptance criteria
      Contract conditions
      Customer role

Project objectives

Project scope definition

Project requirements

Statement of work

Product breakdown structure

Work breakdown structure

Deliverables

Project deliverables

Internal deliverables to inter-related projects

External providers' deliverables

Suppliers' and third-party deliverables

Organization

Stakeholder analysis

Project organisation

Project core team

Roles and responsibilities matrix—RACI Chart

Internal organization

Customer organization

Third-party organization

Project schedule

Project activity network

Project plan Gantt schedule

Milestone schedule

Customer schedule

External providers' schedule

Project resource plan

Resource profiles

Resource loading histograms by profile

Resource commitments

Project cost plan

Cumulative cost curve

Direct costs

Indirect costs

Reserves

Project contracts

Customer

External providers

Risk plan

Risk log

    Analysis
    Mitigation plan
    Contingency plan
    Risk owners
Regulations and standards
    Local and national
    International
    Industry specific
Project control and reporting
    Data collection
    Monitoring
    Evaluation
    Status reporting
    Customer reporting
Supporting plans
    Acceptance
    Change request management
    Commissioning
    Communication
    Configuration management
    Customer training
    Documentation
    Health, safety, and environment
    Implementation
    Integration
    Internal training
    Post-delivery support
    Procurement
    Quality
    Testing
Transition and hand-over

4.7.2 Establishing the Baseline

The baseline project schedule shows the adjusted, acceptable, and approved time-schedule for the project. The resource histograms present the types and profiles of resources that are to be made available and their volume by time-unit.

All entities that will supply resources to the project must establish resource contracts that commit them to make resources available depending on the schedule. This step of the scheduling process is crucial. When many projects are launched or realized concurrently, it is essential that limited resources need to be distributed according to overall company strategies, objectives, and priorities.

Line managers responsible for resources that will be provided to a given project must have a clear picture of their own resource capacity plan. When a resource is assigned to a project, for whatever work effort, it cannot be allocated elsewhere at the same time in such a way to exceed its availability. Resource contracts must therefore be accompanied by a clear understanding of the project's priority with respect to other projects.

During project realization, it is the responsibility of the resource manager to supply the committed resource according to the agreed skill level and/or quality and availability. It is at this stage that the resources, with their specific characteristics, can be assigned to the project activities.

It must be re-emphasized that schedules are for "work to be done." This means that during project realization it may well happen that for a variety of reasons, the original project schedule is no longer valid. A re-scheduling round must then be carried out, which, for the best part, is equivalent to the scheduling rounds described above.

4.7.3 Baseline Approval

The project manager collates the project plan as explained above, and ensures that the management summary is comprehensive and highlights the key contents of the project.

The project plan is officially presented to the sponsor and key stakeholders for review, discussion, and agreement. When any of the three major constraints cannot be met, the project manager presents trade-off options. These will be driven by the scope, cost, or schedule. When necessary, the project plan is forwarded to management prior to the formal baseline approval meeting.

Following management agreement, the sponsor and key stakeholders formally approve the project plan and sign-off. This will authorize the release of funds and the immediate assignments of resources for project start-up.

The project can now proceed to project implementation.

5

Risk Management

5.1 CHAPTER OVERVIEW

This unit covers the project risk management process from the identification of risks to establishing the appropriate risk plans that address these risks and the subsequent monitoring and control.

Central to this chapter is the risk management process, which guides the reader through the steps of identification, assessment, ranking, planning, and control.

5.2 RISK MANAGEMENT—SYNOPSIS

5.2.1 The Risk Management Environment

Project risk management addresses the uncertain events or conditions that, if they occur, have negative or positive effects on the project objectives.

A risk event may have one or more causes and one or more effects. Primarily the effects would be on the major vectors of the triple constraint: scope, cost, and schedule. The effects also extend to cover corporate image and reputation, safety and environmental issues, and the future operability of the project's product(s).

Risk management seeks to protect the project, in fulfilling its objectives in an environment outside its control, by developing proactive and reactive action plans (see Figure 5.1).

Risk management encompasses identifying, analyzing, responding to, and controlling project risks. It aims to minimize the consequences of negative and adverse events and maximize the results of positive events.

FIGURE 5.1
Risk management environment

The process of determining an acceptable level of negative risk during the pursuit of a project, and managing these during project implementation, is necessary to successfully accomplish project objectives.

Risk management must be pursued as an integral part of the project management process. Risks are managed in a concerted effort by the project manager and the project core team members.

5.2.2 Uncertainty and Unknowns

Projects, by definition, are performed in the future, and as a corollary project risks have their origins in the uncertainty and the current levels of the unknowns of the undertaking.

The expected duration of a project presents different time windows of uncertainty, ranging from the immediate future to far future, with corresponding levels of uncertainty, from low and manageable to high and totally speculative. The project manager has to establish an appropriate risk analysis and plan frequency, which corresponds to the project's time frame. The longer the expected duration of the project, the more frequent the cycles of the risk management process.

5.2.3 Assumptions

Assumptions are closely related to the uncertainties and unknowns, and clearly play a major role in risk management. They are by definition virtual facts on which plans are made. They have to be validated for realism. If too many exist, then the project manager must pause the risk management process to determine if further analysis, where possible, is not required to convert assumptions into real facts.

5.2.4 Factors that Govern Control and Lack of Control of Risk Events

Risk events may be under the control and influence of the project manager, or they may depend on external factors and be outside of control. Additionally, risk events vary in their characteristics. They can be:

- Recurring, as they can occur at any time during the project, for example, illness or forced stoppage of a resource
- Unique, in that they can only exist at a prescribed time window of the project, for example, bad weather in the winter causing delays in shipments
- Interdependent, causing a cascading effect, for example, the defect of a provider's product may cause delays in installation, which may lead to penalties for late delivery

The origin of risk events can be internal or external.

5.2.4.1 Internal Risk Events

Internal risk events may be controlled or influenced by the project team using resource assignments and cost/schedule estimates. Examples of internal risk, organized by category, include the following:

Technology	New or untested technology
	Availability of technical expertise
	Customization (design modifications)
Schedule	Resource availability
	Schedule constraints
	Dependencies
Financial	Funding or budget
	Estimate accuracy
Legal	Patent rights or infringement
	Data rights

5.2.4.2 External Risk Events

External risk events are those beyond the control or influence of the project team, such as customer decisions, market shifts, and governing body actions. Examples of external risk, organized by category, include the following:

Unpredictable	Regulatory changes
	Natural hazard
Predictable but uncertain	Market changes
	Inflation

5.2.4.3 Exercise—Influences on Your Current Project

Draft the internal and external influences to your current project and the assumptions that have been made and validated.

5.3 TERMINOLOGY

5.3.1 Risk Event

A risk is the occurrence of a particular set of circumstances and is composed of three basic components:

- A definable event—threat or opportunity
- Probability of occurrence
- Consequence (impact) of occurrence

A risk event is often described as follows:

IF
Conditional statement (allowing probability to be assessed)
THEN
Statement (describing the consequence which can be assessed)

5.3.2 Uncertainty

Uncertainty is a representation of the possible range of values associated with either (1) a future outcome or (2) the lack of knowledge of an existing state. Uncertainty can be expressed as a deterministic quantitative value (i.e., a single number), a qualitative value (high, low, medium, etc.), or as a probability distribution (i.e., a range of quantitative values and the likelihood that any value in the range will occur).

5.3.3 Threats and Opportunities

Threats are risk events that, should they occur, will cause negative effects to the project's objectives. These are defined during the identification step of the project risk management process. Measures need to be taken to reduce their probability and/or impact.

Opportunities are risk events that, should they occur, will bring positive effects to the project's objectives. These are defined during the identification step of the project risk management process. Measures need to be taken to increase their probability and/or impact.

5.3.4 Probability

Probability is the likelihood that a risk event will occur. This can be expressed qualitatively using an ordinal scale: high, medium, or low, or using a cardinal scale: as a single value or as a percentage.

5.3.5 Impact

The impact is the effect and/or consequence on the project if the risk event should occur. This can be expressed qualitatively using an ordinal scale: high, medium, or low; quantitatively in monetary terms; or descriptive of the consequences that can be subsequently estimated and valued in monetary terms.

5.4 VULNERABILITY AND POTENTIAL RISKS

5.4.1 Vulnerability Risks

At any stage of the project's duration, there is a state that describes the known factors with which the project will be managed. These are the known/known (real facts) and the known/unknown (assumptions) segments of the knowledge base. In this context, the project is assessed to determine its current state of vulnerability: the susceptibility to physical or emotional injury or open to attack or damage.

A vulnerability assessment is the process of identifying, quantifying, and prioritizing (or ranking) the vulnerabilities in a system in its current state.

---

1. Vulnerable areas associated with this state of the project
2. Specific potential problems within these vulnerable areas that could have a sufficient negative effect on the project to merit taking action

---

Vulnerability assessments are traditionally performed for systems that include, but are not limited to, nuclear power plants, information technology systems, energy supply systems, water supply systems, transportation systems, and communication systems.

Project risk management can use the same scheme to establish its vulnerability, and decide on the level of confidence the project team has of the project in its current state. Actions can then be taken to address the areas of vulnerability. For example, a project uses a new technology. If the current skill and/or competence level of internal engineers is not adequate, the project is vulnerable since, with the state of knowledge and expertise, the engineers will not be able to perform their tasks. Action then needs to be taken to train and raise the knowledge level of the engineers before pursuing their work, thus decreasing this vulnerability.

Establishing vulnerability will also express acceptable areas confidence of the project. For the example mentioned above, if an external services provider with vast experience and exposure to the new technology is assigned to perform the work, then, in this case, there is little or no issue with vulnerability to the project.

Where risk analysis is principally concerned with investigating the potential risk events, vulnerability analyses, on the other hand, focus on the confidence the team has in the current situation of the project.

5.4.1.1 Establishing Confidence Criteria

The aim of vulnerability risk assessment is to ascertain the level of *confidence* that exists with the current knowledge and facts of the project.

The project manager and the core team members establish a structured list of the project vulnerability areas to assess. The structure will have *domains*, such as technical, financial, schedule, resources, client, contracting, HSE, etc. Each domain is broken down into sets of *characteristics* that give further detail to each of the domains. Characteristics are broken down further into specific *attributes*. Each attribute is given a *confidence feature*—high = 7, medium = 3, and low = 1. Vulnerability analysis consists of determining the level of confidence of each attribute item by mapping the current project state to the appropriate feature.

The characteristics, attributes, and confidence features indicate how the project vulnerability will be determined. The project manager and the core team members establish the criteria for each confidence feature, by clearly stating the description of a particular confidence level. This is an important step, when the team members must be realistic with regards to the requirements of the project and how objectively/subjectively they consider the criteria will protect the project. From the current state and knowledge of the project, the existing information is mapped against each confidence feature and the team ticks the corresponding box:

For example in the *domain* of project procedure under the *characteristic* of structure, an *attribute* titled total project duration could be stated as:

HIGH	MEDIUM	LOW
< 6 months	6–18 months	> 18 months

Using this set of criteria, if the project duration were less than 6 months it would indicate that there is a higher confidence level. However, if the duration were longer than 18 months there would be a low level of confidence for the project.

5.4.1.2 Scoring Vulnerability

As a group effort, and with commonly shared project knowledge, the knowledge and facts of the project are matched to a vulnerability/confidence risk assessment form (as shown in the worked example below). Confidence levels must be determined for each attribute (or characteristic if no attribute has been assigned). A high confidence level will mean a low vulnerability risk.

When all attributes have been analyzed, the total for the characteristic is recorded in the score box. When all characteristics have been analyzed the total for the domain is recorded in the score box. The project total can then be computed as the total of all domains and the overall project confidence can be determined.

5.4.1.3 Establishing Confidence Tertiles for the Project

The first step is to calculate the theoretical maximums for characteristics, domains, and the overall project. The number of attributes is multiplied by 7 to calculate the characteristics total. The domain total is the sum of the characteristics totals, and the overall project theoretical maximum is the sum of the domain totals.

Tertiles are then established by dividing the overall project theoretical maximum into three confidence categories. Assignment of tertiles may be as follows:

The 1st tertile = 1 to 25% of the overall project theoretical maximum
The 2nd tertile = 26 to 50% of the overall project theoretical maximum
The 3rd tertile = 51 to 100% of the overall project heoretical maximum

---

Each tertile corresponds to a *confidence factor*:

Tertile 3	Tertile 2	Tertile 1
HIGH CONFIDENCE	MEDIUM CONFIDENCE	LOW CONFIDENCE

For example, if the overall project theoretical maximum is equal to 364, then the tertiles will be as follows:

1st tertile	1 to 25%	1 to 91
2nd tertile	26 to 50%	92 to 182
3rd tertile	51 to 100%	183 to 364

5.4.1.4 Determining Confidence Levels

When all domains have been assessed, totals are computed. This is done for each characteristic and totalled for each domain. The aggregates per domain give the *overall project score*. This total is then compared to the tertile ranges and is appropriately indicated. For the tertile table in the example just given, if the overall project score is equal to 129, then the project is in the second tertile = medium confidence.

Confidence can be divided into two major categories: *macro* and *micro*. The first relates to the *project as a whole*, and the second to project *attributes*. For example, a project concerning the introduction of a well-known type of component with little or no innovation required may well result as a high macro confidence at the project level; however, the compliance to a new international standard may be a low micro confidence at the characteristic/attribute level.

The *macro project confidence* factor is equivalent to the tertile that corresponds to the overall project score calculated as described previously.

When the macro project confidence factor is low or medium, team members should concentrate on attributes with scores of one (1). These are the low micro confidence risks. The group then needs to establish *preventive action plans* (explained in Section 5.5.6, Risk Response Planning).

It is acceptable for a macro project confidence factor classified as high is manageable in ordinary circumstances to not need further analysis with a view of drafting an action plan. Of course pessimists and/or fearful souls can still perform this analysis, but it might prove to be overkill.

The vulnerability/confidence risk assessment forms are filed in the *risk assessment documentation*. The preventive action plans are inserted in the respective sections of the *risk plan*. The plans are used in subsequent discussions with partners and management, in order to seek agreement and approval.

Results from the preventive action plan are to be incorporated into the project plan and are resourced accordingly.

5.4.1.5 Example—Vulnerability/Confidence

Figure 5.2 shows an example for the domain "subcontractor." If the "financial base" characteristic and "net profitability" attribute of the subcontractor is 7 percent, then the "medium" column is ticked for a score of 3. If the "organizational base" characteristic and "number of offices" attribute of the subcontractor is 1, then the "low" column is ticked for a score of 1.

5.4.2 Potential Risks

Potential risks are at the core of the risk management process. With reference to the knowledge base model, the unknown/known portion describes

Vulnerability/Confidence Risk Assessment

Project Id.	Project Description	Date

Domain	Sub-Contracting	Score	

Characteristic	Financial Base					Score
	Criteria per Confidence Level					

Attribute	High	7	Medium	3	Low	1
No of Years in Business	>5		2–5		<2	
Type of Corporation	Incorporated		Limited Co.		Private	
Annual Sales in US$	>10 Million		2–9 Million		<2 Million	
Net Profitability in %	>10		0–10		Negative	
Current Debt in US$	None		0–5 Million		>5 Million	
......						

Characteristic	Organisational Base					Score
	Criteria per Confidence Level					

Attribute	High	7	Medium	3	Low	1
Number of Offices	>20		2–19		1	
Number of Employees	>50		11–50		<10	
Management Levels	<3		3–6		>7	
International Presence	World-wide		Continental		National	
......						

Characteristic	Experience/Expertise Base					Score
	Criteria per Confidence Level					

Attribute	High	7	Medium	3	Low	1
Number of Consultants	>15		4–15		<3	
Number of Project Managers	>10		1–10		None	
Number of Deals Completed	>10		1–10		None	
Number of References	>5		1–5		None	
Average Years of Experience	>3		1–3		<1	
......						

FIGURE 5.2
Vulnerability risk example

the assumptions that have been established. This is an area that needs to be validated with partners and management. In this context, potential risks relate to the probability that an assumed event will occur. If this event does not occur, it will never be known whether this is because of misjudging the risk or simply by luck. The analysis of potential risks requires assessment of the following:

1. Likely causes of potential problems and their consequences
2. Actions to prevent them from happening
3. Contingent actions that can be taken if preventive actions fail or where they are not undertaken or not possible

Every preventive action that results from the risk assessment has a cost, as it calls for an allocation of resources against future return. This impacts the schedule, and in some cases can modify the technical content of the project.

Assessing potential risks is an exercise of subjectivity and objectivity in looking into the future and applying "What if?" analysis to the project. The assumptions that have been formulated as part of the unknown/known portion of the knowledge base are considered during risk identification. The aim of this activity is to identify the events that may happen that would impact the project objectives.

5.5 THE RISK MANAGEMENT PROCESS

The risk management process is a cyclical and iterative process performed throughout the project development cycle. The risk management process consists of one initial foundation step, and five steps that span from risk identification to risk monitoring and control (see Figure 5.3).

The main emphasis of risk analysis and planning is placed during the planning phase of the project, whereas risk monitoring and control is prominent during the project execution and close-out phases. However, it must be noted that even during the project delivery phases, risk analysis and risk planning are still performed.

The illustration in Figure 5.4 defines the minimum frequency of iterations of the risk management process. During the risk management plan session, the team members would establish an appropriate

FIGURE 5.3
Risk management process

additional number of iterations depending on the duration and complexity of the project, considering the volatility of the environment in which it will be performed.

5.5.1 Risk Management Planning

Risk management planning is the process that establishes the overall approach to determining project risk events, how to address these, and that plans and controls the corresponding actions. (From this point, and for ease of use, the word "risk" refers to a "risk event.")

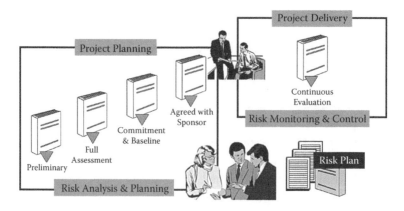

FIGURE 5.4
Risk planning frequency

Risk management planning stresses the importance and necessity of risk documentation and ensures that coherence and visibility of project risks and resulting actions are provided to sponsors and key stakeholders for the provision of corresponding resources and funding for risk management activities.

The risk management planning process should be instituted at the outset of the project planning phase, as it sets the framework for performing the other processes effectively. This is best done by convening participants at a specific meeting. The project manager leads this meeting, in attendance with project core team members and stakeholders. Others in the organization with responsibility for managing the risk planning and execution activities are consulted or invited to meetings when necessary.

The objective of the risk management planning meeting is to agree on the approach for conducting the risk management activities. The project schedule and budget is adjusted to include these activities and their cost. The RACI chart is expanded to include assigned risk responsibilities. Furthermore, existing templates for risk categories and definitions of terms such as levels of risk, probability by type, impact by type, will be tailored to the specific project. The scheme to be followed for qualitative and/or quantitative risk analysis, with the corresponding probability and impact matrix, will be agreed. The results of the risk management planning meeting are included in the risk management plan.

5.5.2 Risk Identification

This step concerns the identification of risks (see Figure 5.5). It is strongly suggested that only risk identification is performed at this stage, and that no attempt is made to analyze or respond to the risks. Proceeding step-by-step will greatly enhance the completeness of the total process.

The data set of identified risks is the first to populate the risk register (see Figure 5.6) for threats or opportunities. Identification can be performed using a structured list by risk categories, with reference to every work package in the WBS, or using the different tools described below.

The project team should ensure that they have a comprehensive knowledge of the key elements of the project:

- Project scope specification
- Product requirements and design (if available)
- Work breakdown structure (WBS)

FIGURE 5.5
Risk identification

- Resource plan and project organization
- Project schedule

Project risk categories provide for a structured approach to developing an extensive list of identified risks. The categories can refer to project components or deliverables, functions, characteristics, functional groups, or any other structure that is advantageous. PMI suggests the structure in Figure 5.7.

Risk Register for **Threats** or **Opportunities**										
Project ID.			**Project Description**					**Date**		

Risk Group	Event N°	Risk Event Description (Threat or Opportunity)	Qualitative Assessment			Quantitative Assessment			Rank	Response Strategy	Risk Plan Number
			Prob.	Impact	Rating	Prob.	Impact	Expected $ Value			

FIGURE 5.6
Risk register

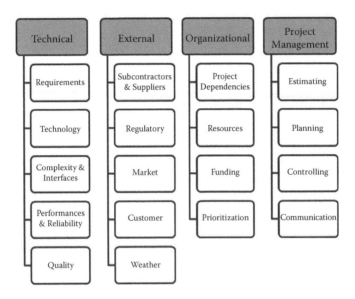

FIGURE 5.7
Risk breakdown structure

The WBS provides a frame of reference by providing a way to ensure that the potential risks in every project area are addressed. This process entails looking at each WBS work package and identifying its associated risks. Some risks, however, may be project related or related to a major activity rather than to a specific work package. These risks also need to be identified. If the large number of work packages makes performing risk identification at that level impractical, the project manager may choose to perform the identification at a higher level.

As risks are identified, they should be listed for full review by all participants. Whenever possible the risks should be referenced to a specific work package or WBS element. During the risk identification process, risks should simply be listed; no judgments should be made regarding their validity.

The risks should be stated clearly using explicit statements describing situations that can be negated or made conditional:

IF
Conditional statement (allowing probability to be assessed)
THEN
Statement (describing the consequence which can be assessed)

For example, IF a delay of one week in supplier delivery occurs during integration, THEN the project delivery schedule will be late by one week, exposing the project to penalties.

5.5.3 Techniques to Identify Risks

5.5.3.1 Checklist

Checklists are usually organized according to the source of the risk. Sources include the project context (the project phase, organizational structure, project stakeholders, project management skills, or socioeconomic influences); the product of the project or technology issues; contractual elements; and other project elements, such as cost, schedule, resources, or procurement.

The selected risks are entered in the risk register.

5.5.3.2 Brainstorming

Risks can also be assessed in a group brainstorming session. A list of potential problems is drafted with the consensus of the team members. This list can be as short or as exhaustive as the team wishes. A "free–wheeling" approach by the group will bring the best result, and the experience of members in the brainstorming session will certainly play a major part in drafting a balanced list. Ground rules must be established by the team, in order not to consider wild and out-of-scope items. During the subsequent risk analysis step, additions to the list will certainly be forthcoming, while previously drafted items may be suppressed.

This method is a group interactive process with the following steps:

- Each project team member is asked to provide input on potential risks for the project at hand.
- Project team members are encouraged to generate new ideas or expand on those of other team members.
- A designated project team member writes down all ideas. None are evaluated or thrown away.
- When all input is exhausted, the group proceeds through the list and discusses the merits of each idea. Some risks are eliminated, and the remaining ones are consolidated and categorized.
- The list of risks is entered in the risk register.

5.5.3.3 Nominal Group Technique

This technique is similar to brainstorming, the difference lies in the approach. The process consists of the following steps:

- Each person on the project team writes down what he or she sees as major risks, with no verbal communication.
- Each team member reads off one item from his or her list, and the items are recorded on a flipchart in front of the group.
- This process continues until all of the individual lists are exhausted.
- Each team member individually categorizes the risks.
- The list of risks is entered in the risk register.

5.5.3.4 Analogy Technique

This technique is based on the acceptance that no project represents a totally new system, and historical data is available. The technique is a combination of checklists and brainstorming (or nominal group technique). It comprises the following steps:

- The type of project activities to be accomplished is reviewed.
- Similar projects or similar pieces of projects are identified.
- Useful and related data from previous project efforts is collected.
- The current project is compared to previous projects and potential risks are identified.

The project team then prepares a list of risks and discusses the validity of each one. Questions are clarified, and the team may eliminate some insignificant risks. After clarification, the team reviews and consolidates the list. After consolidation the risks are categorized into groups. The list of risks is entered in the risk register.

5.5.3.5 Delphi Technique

This technique is a group information-gathering process without personal interaction. The project manager or any of the core team members coordinate:

- The problem is formulated and distributed to predefined experts.
- Experts prepare their opinions on the specific questions or problems and the reasons for those opinions.

- The opinions and reasons are reduced to standard or common statements.
- The experts are shown the aggregated opinions of all the experts surveyed.
- Re–evaluation and further substantiation is requested from the participating experts.
- Iterative feedback continues until no further changes result.
- The experts' final opinions are used to compute a set of median values to represent the expert group opinion.
- The list of risks is entered in the risk register.

5.5.3.6 Exercise—Risk Identification

Identify your current project risks using the techniques listed above and populate the Risk Register for Threats or Opportunities. (See Figure 5.8.)

Risk Register for **Threats** or **Opportunities**			
Project Id.	**Project Description**		**Date**

Risk Group	Event N°	Risk Event Description (Threat or Opportunity)	Qualitative Assessment			Quantitative Assessment			Rank	Response Strategy	Risk Plan Number
			Prob.	Impact	Rating	Prob.	Impact	Expected $ Value			

FIGURE 5.8
Risk register for exercise

FIGURE 5.9
Risk analysis

5.5.4 Risk Assessment/Analysis

In the risk analysis step, the project team evaluates the risks identified in the risk register to determine the probability of risk occurrence, the potential risk impact, and the cumulative impact of multiple risk events (see Figure 5.9). Using this data, the project team can subsequently develop the appropriate risk strategies to address the prioritized threats and opportunities. As for the identification step, at this stage the team should refrain from developing any action plan until the ranking has been performed. If, however, discussions are held, the conclusions should be noted and used during the risk response step.

5.5.4.1 Potential Risks Using the Potential Risk Assessment Form

Risk management documentation for potential risks is produced using the *potential risk assessment form,* with one form for each risk identified in the risk register. The goal of the analysis of each potential risk is to determine the *probability* and the *impact* on the project if the risk occurs. Priorities may be set by the team for the order to follow: this is usually by risk categories.

Before proceeding with the risk assessment/analysis, team members discuss possible causes for the potential risk and record these in the *possible causes* box. The team members then record a detailed description of the consequences and impacts that can be expected in the effects box. Clear descriptions of the possible causes and effects will greatly increase

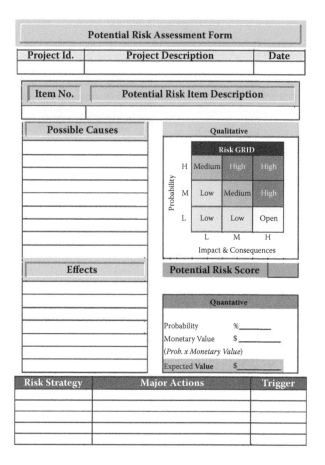

FIGURE 5.10
Potential risk analysis form

the ability of the team to determine the probability of the possible cause and ascertain and/or calculate the impact.

As can be seen in the sample form in Figure 5.10, the form can be used for either a qualitative or quantitative assessment/analysis of the risk.

It should be remembered that risk analysis involves quantifying risks to assess the possible project outcomes. It is also concerned with determining which risk events warrant response. This analysis is complicated by a number of factors:

- A single risk event may cause multiple effects.
- Opportunity for one stakeholder (reduced cost) may be risk to another (reduced profit).

- Not all risks are quantifiable; some can be defined only qualitatively.
- The mathematical techniques used may give the false impression of precision and reliability.

5.5.4.2 *Special Consideration in Risk Assessment/Analysis*

Established rating system guidelines or metrics should be used to assess risks. The project manager and the core team members should determine the needed metrics. Since some risks are difficult to quantify, agreed-upon criteria for qualitative terms such as "high," "moderate," and "low," or any variations thereof, may need to be established. In addition, a narrative method may be used, in which case the process for assigning a quantitative value or a qualitative term to the probability of occurrence and the impact of occurrence must be determined. It is important that these risk factors be applied on all projects to avoid misinterpretation.

The project team analyzes each risk, one at a time, and establishes a risk factor for the *probability* of occurrence. This step can be performed separately from, or concurrently with, determining the impact of occurrence. This method should be the preferred one. Whenever possible the project team should try to quantify the probability of occurrence.

In some cases the project team does have enough information to assign a numeric probability to the risk. Simple qualitative methods can then be used to assign risk probability factors of high, moderate, or low, based on the subjective assessment of the risk by the team.

The *impact* is the risk-event value, an estimate of the gain or loss that is incurred if the risk event occurs. The project manager and the core team members should review one risk at a time and determine the impact on all related areas if the event occurs. The impact could affect the cost, schedule, or scope quality, or a combination thereof. The impact analysis can be extended to cover company reputation or HSE. The team should define not only how great the impact is, but also which project element is affected the most.

More comprehensive quantitative methods are preferred if sufficient information is available. Alternatively, simple qualitative methods can be used to determine the impact of occurrence by assigning high, moderate, or low ratings based on subjective assessment of the risk.

The project manager and the core team members should try to quantify the risk impact whenever possible. If, however, the qualitative method is used, it should be used in conjunction with a well-defined possible causes and effects description.

The three major assessment/analysis techniques are explained in the following sections.

5.5.4.3 Narrative Risk Assessment/Analysis

Risks are described in a narrative form. The intention is to describe the risk and its potential impact as clearly as possible. This is the weakest way to assess/analyze risks. It is, however, acceptable when the project team needs to air their comments on the risks, especially during the early stages of the project.

5.5.4.4 Qualitative Risk Assessment/Analysis

In the qualitative risk assessment/analysis, risk is expressed through a rating system using adjectives (high, medium, and low) to denote order. In qualitative analysis the specific data to make a quantitative determination are not available; therefore, a more subjective scale is used employing language such as "there is a moderate probability that this will happen" and "if it happens, it will have a high impact." The qualitative rating must be established and agreed upon prior to qualitative assessment/analysis.

Proceeding with each risk, and using the corresponding risk assessment/analysis form, team members post their opinion on the qualitative risk grid. The grid is a 3 × 3 matrix having probability and impact as the axes, and each axis is divided into high, medium, or low. Each team member's opinion is recorded as a cross in the relevant boxes. From this a cluster of crosses will indicate the team's opinion. The location of the cluster will determine the *potential risk score* as high, medium, low, or open (see Figure 5.11). If the cluster is in the open box, this indicates that the probability is low and the impact high. This is usually the case for human factor risks, critical mission risks, and disasters of many kinds. After conferring on the issue, the team then decides on changing the potential risk score to high, medium, or low. The score will subsequently serve to determine the ranking and appropriate risk response actions.

5.5.4.5 Quantitative Risk Assessment/Analysis

Quantitative risk assessment/analysis is the preferred technique, as it offers a quantifiable monetary value to the risk. However, this type of

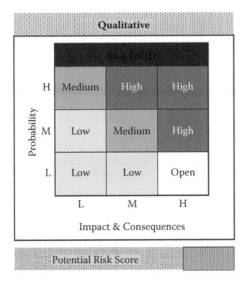

FIGURE 5.11
Probability/Impact risk grid

analysis requires sound quantitative data to determine both probability and impact, and may necessitate a substantial effort from the team to derive the monetary value.

The probability can be determined as a percentage by mathematical extrapolation. When this cannot be done, then a qualitative probability can be converted to an integer. The usual approach is to assign a percentage as follows: High = 85 percent; Medium = 50 percent, and Low = 15 percent.

The impact monetary value is determined by estimating the effects to scope and schedule, as well as the associated costs due to penalties or mobilization resources. Estimates have also to be made for impacts to reputation and HSE issues. These two latter impacts may prove difficult to assess (see Figure 5.12).

Project scope impacts result from factors such as:

- Increased requirements
- Reduced work effort
- Product reliability
- Quality level of deliverable
- Product safety

Risk Register for **Threats** or **Opportunities**		
Project Id.	**Project Description**	**Date**

Risk Group	Event N°	Risk Event Description (Threat or Opportunity)	Qualitative Assessment			Quantitative Assessment			Rank	Response Strategy	Risk Plan Number
			Prob.	Impact	Rating	Prob.	Impact	Expected $ Value			
Schedule	TH01	If product delivery is delayed 1 week, then project schedule is delayed by 2 weeks with $50 000 penalty per week				40	100K	40K			
	TH02	If due to complexity, system integration is delayed by 2 weeks, then resources immobilized will resources immobilized will cost an extra $100 000 per week with no impact to overall schedule				10	200K	20K			
....	...	....									
Schedule	OP01	If higher skilled resources are assigned to testing, then 1 week reduction on the schedule and $40 000 can gained				10	40K	8K			
	OP02	If commissioning is completed before the month of July, then 3 weeks can be gained on the schedule with a cost reduction of $150 000				30	150K	45K			

FIGURE 5.12

Risk register example

Project cost impact drivers cover such factors as:

- Level of effort
- Task duration
- Human resource
- Materials
- Equipment and tools

Project schedule impacts are analyzed for:

- Duration expansions
- Resource shortages
- Other delays

Quantitative values should be assigned when there is a reasonable degree of certainty or confidence in the validity of those quantities. Trying to guess and applying quantitative value to events that are unpredictable in terms of probability of occurrence and impact could convey a false impression of accuracy and, therefore, be misleading.

Quantitative	
Probability	%________
Monetary Value	$ __________
(Prob. x Monetary Value)	
Expected **Value**	$__________

FIGURE 5.13
Expected value calculation

The result of the quantitative probability/impact analysis is posted in the quantitative box in the potential risk assessment form. The statistical expected value is calculated as the product of the probability and the monetary value (see Figure 5.13). For example, a probability of 25 percent for a monetary value of $80,000 results in an expected value of $20,000.

5.5.4.6 Expected Monetary Value (EMV)

The expected monetary value (EMV) analysis is a statistical concept that calculates the average outcome when the future includes scenarios that may or may not happen (i.e., analysis under uncertainty). The EMV of opportunities will generally be expressed as positive values, while those of risks will be negative. EMV is calculated by multiplying the value of each possible outcome by its probability of occurrence, and adding them together. A common use of this type of analysis is in decision tree analysis (explained below in Section 5.5.4.7).

The EMV for threats is called exposure, whereas for opportunities, expected value is called leverage.

The EMV also allows the project manager to determine a best- and a worst-case scenario. From this the project-level expected value can be derived, adapting the original project cost by adding threat exposure and subtracting opportunity leverage.

With reference to the example we are exploring, Figure 5.14 shows the calculations. The current project cost is $1 M. The best case, most likely EV, and the worst case are each entered for the threats and opportunities (note that since this is a cost calculation, the opportunity totals are negative). The EMV result indicates the range from best case to worst case to be $810 K to $1,300 K. By modifying the initial project cost with the threat exposure and the opportunity leverage, the statistical EV of the project can be established at $1,002 K.

Expected Monetary Value		
Project Id.	**Project Description**	**Date**

Risks	Event N°		Prob. %	K$ Value	Best Case	Most Likely EV	Worst Case
		Base Project Cost		1 000	1 000	1 000	1 000
Threats	TH01	If project delivery is delayed 1 week, then project schedule is delayed by 2 weeks with $50 000 penalty per week	40	100	0	40	100
	TH02	If due to complexity, system integration is delayed by 2 weeks, then resources immobilized will cost an extra $100 000 per week with no impact to overall schedule	10	200	0	20	200
		Threats Totals			0	60	300
Opportunities	OP01	If higher skilled resources are assigned to testing, then 1 week reduction on the schedule and $40 000 can gained	20	−40	40	−8	0
	OP02	If commissioning is completed before the month of July, then 3 weeks can be gained on the schedule with a cost reduction of $150 000	30	150	−150	−50	0
		Opportunities Totals			−190	−58	
		Expected Monetary Value Result			810	1 002	1 300

FIGURE 5.14
Expected monetary value—worked example

5.5.4.7 Decision Trees

Decision trees are used to determine the implications of the available choices, or impacts, and possible scenarios for a risk under consideration. A decision tree integrates the cost of the available choices, the probabilities of each possible scenario, and the total EV of each alternative logical path. The decision tree calculation provides the EMV for each alternative, from which the project manager and the core team members can decide on the most appropriate path to take.

To build a decision tree, follow the following technique:

- Decision nodes are represented by boxes; outcomes or events are represented by circles, end of path by a triangle.
- Proceeding from left to right, the primary decision is entered in the tree.
- All possible scenarios are represented by paths.
- Probabilities are assigned to all path segments leading from events.
- Expected value for each segment is determined.
- Proceeding from right to left, expected values of all path segments are added leading to a decision node.
- The most advantageous path is then established.

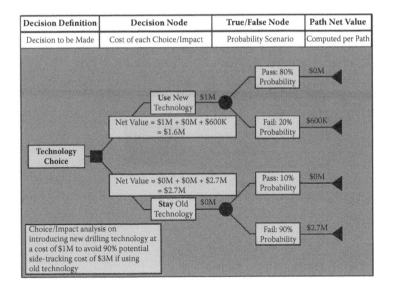

Decision Definition	Decision Node	True/False Node	Path Net Value
Decision to be Made	Cost of each Choice/Impact	Probability Scenario	Computed per Path

FIGURE 5.15
Decision tree—worked example

In the example in Figure 5.15, a decision tree is constructed to determine if the investment in a new drilling technology, at an additional cost of $1 M, can dramatically reduce the probability of a side-track (drilling off the original well path), if the old and unreliable technology is maintained. Side-tracking is estimated to cost an additional $3 M and there is a 90 percent probability of this occurrence with the old technology. The new and more reliable technology is estimated to have a 20 percent probability of forcing a side-track.

By calculating the totals along the different paths, the investment will cost a net value of $1.6 M, whereas maintaining the old technology will cost a net value of $2.7 M.

5.5.4.8 Exercise—Risk Analysis

> Select at least ten threats and three opportunities for your current project and develop qualitative assessment.
> If you possess sufficient information, attempt a quantitative analysis and an Expected Value calculation to determine total best, most likely EV, and worst cases.

FIGURE 5.16
Risk ranking

5.5.5 Risk Ranking/Prioritizing

The step of ranking and prioritizing risks is performed on the data collected following both qualitative and quantitative risk assessments (see Figure 5.16). The ranking criteria must be previously agreed and is recorded in the risk management plan.

In many projects the number of risks identified may be overwhelming. The project team cannot effectively track and manage all of them. Prioritizing risks will focus the team's effort on the most critical ones, those risks that have the highest impacts.

Although the project team may elect not to focus on some risks determined to have a moderate or low overall impact, these risks must be tracked and monitored because their probability of occurrence and their impact may change as the project progresses. What was initially considered a low risk may suddenly become a risk with high overall impact.

When quantitative assessment is the chosen technique, then the ranking is usually performed on the EMV. At times, and depending on management propositions, the criteria can either be on the probability or on the full impact value.

On occasions, a weight factor may be applied to further quantify the risk, as represented in the following formula: Impact×Probability ×Weight Factor = Overall Risk Impact.

The process, however, is not that simple if the qualitative or narrative method, or a combination of all three methods, has been used to determine the probability and the impact. In this case the assessment is subjective, and the only method that can be used is the expert judgment method, in which subject matter experts express their opinions about the probability, impact, and overall risk associated with a given risk event.

When using the result of the qualitative assessment as the criteria, threats (and opportunities in another pass) are sorted on the potential risk score: H, M, or L.

5.5.5.1 Ranking by Paired Analysis

The paired analysis technique is used when the result of the initial ranking cannot differentiate between risks of the same priority. This is often the case for risks assessed qualitatively that have the same potential risk score: H, M, or L. The technique can also be used when a small number of risks have to be prioritized. This number is usually less than ten.

The project manager and the core project team members, in a group consisting of five to seven individuals, will compare the relative importance of any two risks and vote for the one they consider to be of higher importance. Each risk is compared, one by one, to all the other risks. The voting result is recorded. At the end of the analysis, aggregates are made for each risk, and the ranking can then be established.

In the example in Figure 5.17, a team of five individuals vote to rank six risks: T01 is compared to T02, then to T03 and so on. T02 is compared to T03 and so on. The results of the vote between the two risks are recorded. In this example, comparing T01 to T02, the voting is 1 for T01 and 4 for T02.

Risk Number	Risk Description
T01	AAAAAAAAAA
T02	BBBBBBBBBB
T03	CCCCCCCCCC
T04	DDDDDDDDDD
T05	EEEEEEEEEE
T06	FFFFFFFFFF

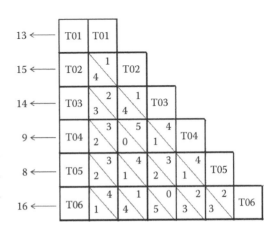

FIGURE 5.17
Paired analysis—worked example

When the voting is complete, aggregates are made for each risk. In this example, T06 has the highest ranking, and T05 has the lowest.

5.5.6 Risk Response Planning

Developing the risk response, referred to as "risk management strategies," entails identifying various risk strategies for each risk, evaluating the effectiveness of each option, and selecting the best approach (see Figure 5.18).

For every risk, several strategies or alternatives may apply. The project manager and the team members must identify these alternatives, evaluate their individual merits, and select the one that offers the best solution.

The risk response process is established during the risk management planning step. Participants in risk response development must be knowledgeable of this process. These participants are usually drawn from the project team but may be a subgroup of the team or may include subject matter experts external to the team. Two primary approaches are commonly used:

- The project team can work as a group in developing strategies for each risk, using the risk analysis worksheets.
- The project manager can divide the risks among the team members based on their expertise. Each team member works independently on the risks assigned, determines the options available, and selects the most appropriate one.

The project team should consider several strategies for each risk contained in its prioritized list, evaluate the overall impact of each strategy, and select the one that provides the best approach. The project team should

FIGURE 5.18
Risk response

recognize that the implementation of a preventative strategy may create new risks. Each strategy should be evaluated based on its merit and how it affects other elements of the project.

The potential risk assessment forms are used to develop the strategies and the corresponding actions (see Figure 5.19). The resulting preventative or contingency action plans will subsequently be inserted in the respective sections of the *risk plan*. These will be discussed with partners and management, to seek agreement and approval.

Results from any *preventive action plan* are to be incorporated into the project plan and are resourced accordingly. Triggers/controls for the *contingency action plan* are handled in the same way and are noted in the Risk Register (see Figure 5.20).

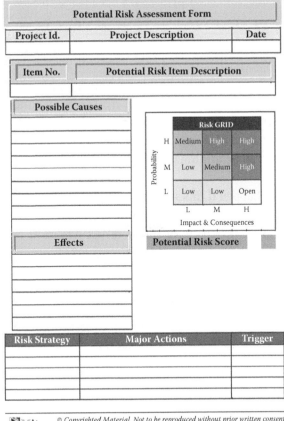

FIGURE 5.19
Potential risk form

Risk Register for **Threats** or **Opportunities**		
Project Id.	**Project Description**	**Date**

Risk Group	Event N°	Risk Event Description (Threat or Opportunity)	Qualitative Assessment			Quantitative Assessment			Rank	Response Strategy	Risk Plan Number
			Prob.	Impact	Rating	Prob.	Impact	Expected $ Value			

FIGURE 5.20
Risk register for threats and opportunities

Implementing some risk strategies can create new and unexpected risks. The project team evaluates them to determine the expected result and the potential impact each may have on other activities, as well as the program. Additional risk can be minimized by asking the following questions:

- What will happen if this strategy is implemented?
- Will any other work package or project element be affected?
- What will be the risk probability of occurrence if this strategy is implemented?
- Is this strategy the best possible one?
- What are the additional cost/schedule impacts if other work packages are affected?

5.5.6.1 Risk Response Planning for Threats

The four basic risk strategies for threats are (from *PMBOK® Guide*, pp. 303, 304):

- **Avoid:** risk avoidance involves changing the project management plan to eliminate the threat posed by an adverse risk, to isolate the project objectives from the risk's impact, or to relax the objective that is in jeopardy, such as extending the schedule or reducing scope. Some risks that arise early in the project can be avoided by clarifying requirements, obtaining information, improving communication, or acquiring expertise.

- **Transfer:** risk transference requires shifting the negative impact of a threat, along with ownership of the response, to a third party. Transferring the risk simply gives another party responsibility for its management; it does not eliminate it. Transferring liability for risk is most effective in dealing with financial risk exposure. Risk transference nearly always involves payment of a risk premium to the party taking on the risk. Transference tools can be quite diverse and include, but are not limited to, the use of insurance, performance bonds, warranties, guarantees, etc. Contracts may be used to transfer liability for specified risks to another party. In many cases, use of a cost-type contract may transfer the cost risk to the buyer, while a fixed-price contract may transfer risk to the seller, if the project's design is stable.

- **Mitigate:** risk mitigation is a preventive action strategy and implies a reduction in the probability and/or impact of an adverse risk event to an acceptable threshold. Taking early action to reduce the probability and/or impact of a risk occurring on the project is often more effective than trying to repair the damage after the risk has occurred. Adopting less complex processes, conducting more tests, or choosing a more stable supplier are examples of mitigation actions. Mitigation may require prototype development to reduce the risk of scaling up from a bench-scale model of a process or product. Where it is not possible to reduce probability, a mitigation response might address the risk impact by targeting linkages that determine the severity. For example, designing redundancy into a subsystem may reduce the impact from a failure of the original component.

- **Acceptance:** a strategy that is adopted because it is seldom possible to eliminate all risk from a project. This strategy indicates that the project team has decided not to change the project management plan to deal with a risk, or is unable to identify any other suitable response strategy. It may be adopted for either threats or opportunities. This strategy can be either passive or active. Passive acceptance requires no action, leaving the project team to deal with the threats or opportunities as they occur. The most common active acceptance strategy is to establish a contingency reserve, including amounts of time, money, or resources to handle known or even sometimes potential, unknown threats or opportunities.

The table here lists possible actions associated with each strategy.

Avoid	Reorganize the project structure
	Negotiate removal of risk-bearing performance requirements with the customer
	Do not launch the project if too much risk is present
Transfer	Contract out certain project tasks to professionals and subcontractors
	Request that the customer perform some tasks that represent a risk to the provider
	Buy insurance
Mitigate	Develop plans to deal with potential risk events
	Closely monitor and control the cost and schedule for the risk event
	Monitor and manage the technical performance of tasks having a high risk factor
	Establish firm resource commitments for tasks at risk
	Negotiate and implement changes to minimize or remove risk events
	Establish a management reserve to handle potential cost overrun risk
Acceptance	Practice proactive acceptance by developing a contingency plan and executing it should the risk event occur
	Price the mitigation contingency based on overall risk impact

5.5.6.2 Risk Response Planning for Opportunities

The four basic risk strategies for opportunities are (from *PMBOK® Guide*, pp. 303, 304):

- **Acceptance**: as explained for threats in the previous section.
- **Exploit:** this strategy may be selected for risks with positive impacts where the organization wishes to ensure that the opportunity is realized. This strategy seeks to eliminate the uncertainty associated with a particular upside risk by making the opportunity definitely happen. Directly exploiting responses include assigning more talented resources to the project to reduce the time to completion, or to provide better quality than originally planned.
- **Share:** sharing a positive risk involves allocating ownership to a third party who is best able to capture the opportunity for the

benefit of the project. Examples of sharing actions include forming risk-sharing partnerships, teams, special-purpose companies, or joint ventures, which can be established with the express purpose of managing opportunities.

- **Enhance:** this strategy modifies the size of an opportunity by increasing probability and/or positive impacts, and by identifying and maximizing key drivers of these positive-impact risks. Seeking to facilitate or strengthen the cause of the opportunity, and proactively targeting and reinforcing its trigger conditions, might increase probability. Impact drivers can also be targeted, seeking to increase the project's susceptibility to the opportunity.

5.5.6.3 Risk Reserves

To complete the response planning process, the project manager and the team establish risk reserves. These need to be discussed, agreed, and approved by the project sponsor and the key stakeholders. Reserves are additional funds or time provided for in the project management plan to respond to cost, schedule, or performance risk. Reserves may be included in the total project budget. However, whatever funds are reserved, they need to be authorized, so that they are made available as and when the risk is realized. Seeking additional funding once the risk occurs during project implementation will need to follow a management approval process. This can at times be a lengthy process, at a time when the project manager needs to resolve the risk occurrence as quickly as possible.

Reserves are intended only for risks within the current scope of the project and do not cover the funding of eventual change requests. The reserves are *not* part of EVM calculations.

- **Management reserves:** these are funds reserved for unplanned changes to project scope and cost. The funds are intended to authorize additional cost and time needed to reduce the impact of missed performance objectives that were impossible to plan. Management reserves are used to address "unknown/unknowns."
- **Contingency reserves:** fund allowances for established contingency plans that are triggered as and when the risk is realized. The resulting applied actions from the contingency plan modify the cost, schedule, or performance objectives.

FIGURE 5.21
Risk monitoring

- **Estimation contingency reserves:** these are funds reserved to cover a known range of estimation errors for cost and schedule. These funds can be specifically allocated to these reserves, or can be integrated into the management reserves.

5.5.7 Risk Monitoring and Control

Risk monitoring and control is the process of tracking identified risks, reassessing existing risks, monitoring trigger conditions for contingency plans, identifying, analyzing, and planning for new risks, and reviewing the execution of risk responses while evaluating their effectiveness (see Figure 5.21). Project assumptions are validated and/or modified, and the use of reserve funds is incorporated into the overall project cost and schedule.

The risk monitoring and control process applies variance and trend analysis, from performance data generated and collected during project execution. Risk monitoring and control is an ongoing process throughout the life of the project.

5.5.7.1 Early Warning Mechanism

During the development of the contingency plans, triggers are set and recorded. Many techniques can be used to detect the deviations from the project plan which action these triggers. This may be simple reviews integrated during the standard project control process, or another scheme that highlights the evolution of the conditions upon which the trigger is tripped. The most common is the use of a traffic light system of green, amber, and red. During project risk review meetings, or from the risk owner source, the related light is set to the corresponding state of evolution of the risk.

5.5.7.2 Risk Tracking and Status

Any change to the original project risk plan, because of a triggered contingency plan, an additional risk, or even a dropped risk, must be reflected in the project plan and schedule.

It is strongly suggested that the change request management process is invoked to record the changes on the appropriate change request form (discussed in Chapter 7, Project Implementation). It is imperative for the project manager to substantiate variations of this nature, and to include these in the final end of project report.

5.5.7.3 Risk Evaluation

The project manager and the team members must institute a continuous risk evaluation mechanism. Areas of evaluation can cover the following:

- Have mitigation risk actions been effective?
- Have mitigation risks been sufficiently contained and are further actions needed?
- Have triggered contingency plans been effective?
- Are any passively accepted risks that have materialized tolerated?

5.5.8 Risk Documentation

The *risk management plan* documents the procedures that are used to manage risks throughout the project. The plan incorporates the risk identification and analysis data generated in the identification, assessment, and response planning steps. The plan should cover who is responsible for managing various areas of risk, how the initial identification and analysis output are maintained, how contingency plans are implemented, and whether management funding is allocated.

The risk management plan is a supporting element of the project plan. It may be highly detailed or broadly framed, depending on the needs of the project. However, the project's risk management plan should contain at least the following elements:

- A summary listing of all of the risks identified in the risk register
- All of the completed risk assessment/analysis forms
- A detailed plan (mitigation or enhance strategy) for each major risk to be tracked
- A summary of all contingency plans and their respective trigger conditions
- Identification of the risk owners for tracking or resolving each risk
- A risk management plan review-and-update schedule
- A risk-result documentation process

5.5.9 Documentation Process

The project manager reviews the risk assessment/analysis forms and risk register to confirm that all of the project risks have been identified and addressed (see Figure 5.22).

The project team develops the risk management plan. The plan includes the following:

- **Risk Summary**—An overview of the significant project risks and the risk responses that have been developed
- **Risk Assessments**—A detailed evaluation of all project risks that reference the risk assessment/analysis forms and risk register

FIGURE 5.22
Risk management documentation

- **Risk Management Approach**—The risk management strategy that is applied to the project in terms of monitoring risk triggers, evaluating the effectiveness and accuracy of the risk management plan, and the process for following up on risk events to ensure a satisfactory conclusion.

In preparing the risk management plan, the project team selects the best strategies for each of the major risks and incorporates them into the risk management plan and/or supporting documentation. These strategies should include the implementation details along with pricing and schedule contingencies.

Preventive strategies should be entered into the WBS as specific tasks, and the resources should be assigned as for any other task. The same applies to the pricing and schedule where risk contingencies should be clearly identified.

Reactive strategies, with information about the risk event, should be documented in the risk management plan to be referenced and implemented if the risk event occurs. Reactive strategies should be tied to milestones that trigger the specific strategy if a given risk event takes place.

The risk management plan essentially ensures an adequate risk response for the project. Its importance is highlighted each time a response to a risk is needed. The project team uses this critical document throughout the project to demonstrate awareness and preparation for managing risks.

Once the risk management plan has been completed, it is reviewed and approved to ensure that all project risks have been identified and adequately addressed. Depending on the scope of the project, this review may be extended to include the customer and any significant project participants.

5.6 GLOSSARY OF TERMS

Contingency plan: "A planned and documented set of actions to be taken in response to a risk event when it has occurred." Usually related to threats rather than opportunities and implemented if proactive response plans have not been identified or have failed to prevent occurrence of the event and/or its impact. The cost of these reactive responses is met from contingency.

Decision tree: A branched diagram consisting of a sequence of nodes (representing decisions or uncertainties) and outcomes associated with each branch. The purpose of a decision tree is to define the set of scenarios and the sequence of events that guide the evaluation of risk and return.

Event: Occurrence of a particular set of circumstances.

Expected value: "The weighted average outcome using the probabilities as weights." For decisions involving uncertainty, the concept of expected value provides a way of selecting the best course of action and of forecasting portfolio level performance.

Note: Do not confuse "expected value" with "most likely." If, for example, an event has a 90 percent chance of yielding $10 but a 10 percent chance of yielding zero, then the expected value is $9, although in an individual case this outcome is impossible and the most likely outcome is $10.

Manageability: An expression of the ability to mitigate a threat or leverage an opportunity, demonstrated by creating response plans that are expected to be effective.

Mitigation: A proactive risk response to a threat.

Monte Carlo simulation: A statistical analysis process that takes a random sample from each of the input probability distributions and combines the sampled values according to the equations in a model to produce an overall probability distribution for the output.

Proactive risk response: An action or set of actions to reduce the probability or impact of a threat (or delay its occurrence), or increase the probability or impact of an opportunity (or bring forward its occurrence). Proactive risk responses, if approved, are carried out in advance of the occurrence of the risk. They are funded from the project budget.

Reactive risk response: An action or set of actions to be taken after a risk event has occurred (as defined by the trigger condition) in order to reduce or address the effect of the threat, or maximize the effect of the opportunity. The cost of reactive risk responses is met from contingency (unallocated provision). More usually applied to threats, and detailed within a contingency plan.

Risk: An event (or set of circumstances) that, should it occur, would have a material effect, positive or negative, on the final value of a project. Risks with a positive impact are called opportunities while those with a negative impact are called threats. Risks may be characterized by (1) their probability of occurrence, and (2) the impact of occurrence on the project value (expressed in monetary terms where possible). Either or both of these parameters may be represented by a probability distribution where the true value is uncertain.

Risk register: A tool (spreadsheet or database) containing all the risks identified for a project, along with a description of each risk and a documentation of information relevant to the ownership, assessment and response of each risk.

Risk response: Action taken to reduce the probability of a threat arising or to reduce its impact if it were to arise. For an opportunity, the response aims to increase the probability of it arising and to increase its beneficial impact. Proactive risk responses (mitigations) are funded from within the project budget, while reactive responses (interventions) are funded from contingency.

Trigger condition: A definition of the circumstances in which a risk is deemed to have occurred, or upon which a reactive response will be initiated.

6

Procurement Management

6.1 CHAPTER OVERVIEW

This chapter describes and illustrates procurement and contract management for projects. It covers determining procurement requirements, establishing agreed procurement processes, soliciting products and services, conducting contracting and procurement activities, and managing finalization processes.

The project manager is pivotal in procurement management. This chapter will explain in detail the steps of the procurement activities, as well as the roles and responsibilities of subject-matter experts in these activities.

In this chapter, the words "purchase" and "procure" are used interchangeably. The external organization fulfilling its contract of procurement will be named variably as the provider, vendor, seller, or outsourcer. Use of products and services procured will also cover items such as goods, merchandise, works, and fees.

Please note that this chapter uses the word solicit, and its variants, in its appropriate sense of a request, unlike the PMBOK that refrains from using a normal English word, because it considers the word to have other connotations.

6.2 PROJECT PROCUREMENT MANAGEMENT SYNOPSIS

Project procurement management (PPM) is part of the project management process in which products or services are acquired or purchased from outside the organization in order to complete the task or project. Project procurements for products or services are bound under a legal contract.

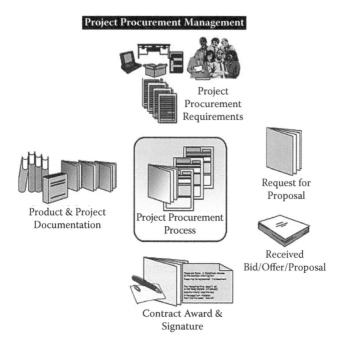

FIGURE 6.1
Project Procurement Management framework

PPM spans a variety of tasks, initiated by a planning step, during which product or services to acquire or purchase are identified and decided—this establishes the PPM plan to be utilized. Requests are made to provider organizations that respond with a bid or proposal. Following vendor selection and award, a contract is drafted that provides a legal document of the exchange. This sets in motion the procurement of the product or service. A pivotal element of PPM is the project management relationship between both the buyer and seller organizations and their representatives via the contract (see Figure 6.1).

Project objectives are reviewed to ensure the acquisition is aligned to the stated objectives, resources necessary for the acquisition are identified and the contract type needed to secure the acquisition is determined.

Soliciting responses is made by issuing a formal document, which describes in detail the products or services requested, for example a request for proposal.

From the responses and proposals received, specific vendors are identified and placed on a qualified sellers list. Selected vendors are considered qualified according to their ability to provide the products or services within the constraints of the project, their interest in providing these, and the acceptability of their bids. Proposals are evaluated in order to determine the best vendor to deliver the products or services. Following final seller selection, contracts are negotiated.

Contract administration ensures that the obligations, responsibilities, and performance goals of the vendor are fulfilled. Contract changes should be controlled and documented to prevent unnecessary legal claims.

Once the contract is complete, the contract is audited to make certain all terms of the contract were accomplished. Contract closure involves evaluating the performance of vendor and documenting any lessons learned in executing the contract.

6.2.1 Procurement Roles and Responsibilities

The project manager may be responsible for the actual procurement of the services or products needed to develop and implement the project, or may be directing these activities through a contracting or procurement individual who will act as an administrator. Procurement may be centralized in the organization or decentralized within the performing structure. All procurement activities have a significant impact on the project budget and schedule, and must be integrated into the overall project plan and project schedule.

The following responsibilities should be delegated to the major roles, listed below, to ensure that the project procurement tasks are performed by the appropriate subject matter expert.

6.2.1.1 Project Manager

The project manager should be conversant in the project scope of work and the technical/business objectives.

The project manager also should be assigned to the project as early as possible to help ensure that the contract, its terms and conditions, and the scope of work are appropriate to the scope of the project.

- Identify risk and incorporate mitigation and allocation of risks into the contract
- Participate in the development of the RFP, proposal evaluation, and contract negotiation
- Adapt the contract to the need of the project

- Include the schedule for completion of the procurement process into the schedule for the project
- Protect and manage the relationship with the vendor
- Responsible for project termination

6.2.1.2 Project Contract Administrator

The contract administrator is responsible for compliance of the vendor to the contract's terms and conditions, and ensures that the final product or service complies with the statement of work specifications.

- Monitor product or service delivery
- Ensure adherence to quality
- Perform contract change management
- Establish warranties
- Resolve contract disputes/breach
- Manage payment schedules to ensure that contract terms and conditions are met before funds are disbursed to vendor
- Handle contract closeout

6.2.1.3 Project Management Office (PMO)

The PMO is conversant in the project scope of work and the technical/business objectives.

- Assist project manager in identifying and assessing contract risks and in developing and incorporating mitigation plans into the project plan
- Ensure that appropriate expert knowledge is available for make/buy decisions
- Benchmark procurement expectations from past and similar projects
- Assist project team with structured vendor performance reviews
- Assist the project team prepare for proposal evaluation and contract negotiations
- Support the project team to adapt the contract to the needs of the project

6.3 PROJECT PROCUREMENT PLANNING

A project procurement management plan (PPMP) is an indispensable component of the project. It is required when the decision to purchase has been made. The project manager is responsible for delivering the PPMP.

It is essential and imperative that the project management team and/or the project management team leader implement an effective and succinct plan when it comes to the various procurement management activities throughout the project's life cycle. The project procurement management plan refers to the instituted plan that describes the overall approach of the procurement process for the required products and services and states the detailed steps to follow, the documentation to draft, the contract award methodology and how contracts will be administered and closed. The project procurement management plan should be implemented and developed as early in the project life cycle as possible to assure that, to the extent possible, the procurement process is consistent throughout. However, in some cases, the plan may be altered once the project begins, particularly if budgetary reasons dictate.

6.3.1 The Project Procurement Plan

The project procurement plan is established in the initial phase of the project life cycle. At this juncture in the project procurement planning steps, the plan is high level; however, in most cases the decision to make or buy product has been established. The project procurement plan parallels the phases of the project plan. Most of the procurement efforts, however, occur in the project planning phase. In some cases, decisions to procure additional products and services may be identified through the planning processes. In all cases, contracts are awarded before a project is approved to proceed to the execution phase, when contract administration is initiated.

The detailed project procurement plan is established by the project manager (see Figure 6.2). **Specialist procurement staff can help to prepare the contents. The project procurement plan must be monitored and updated to accommodate any changes enforced by changes in the marketplace and those due to internal policy.**

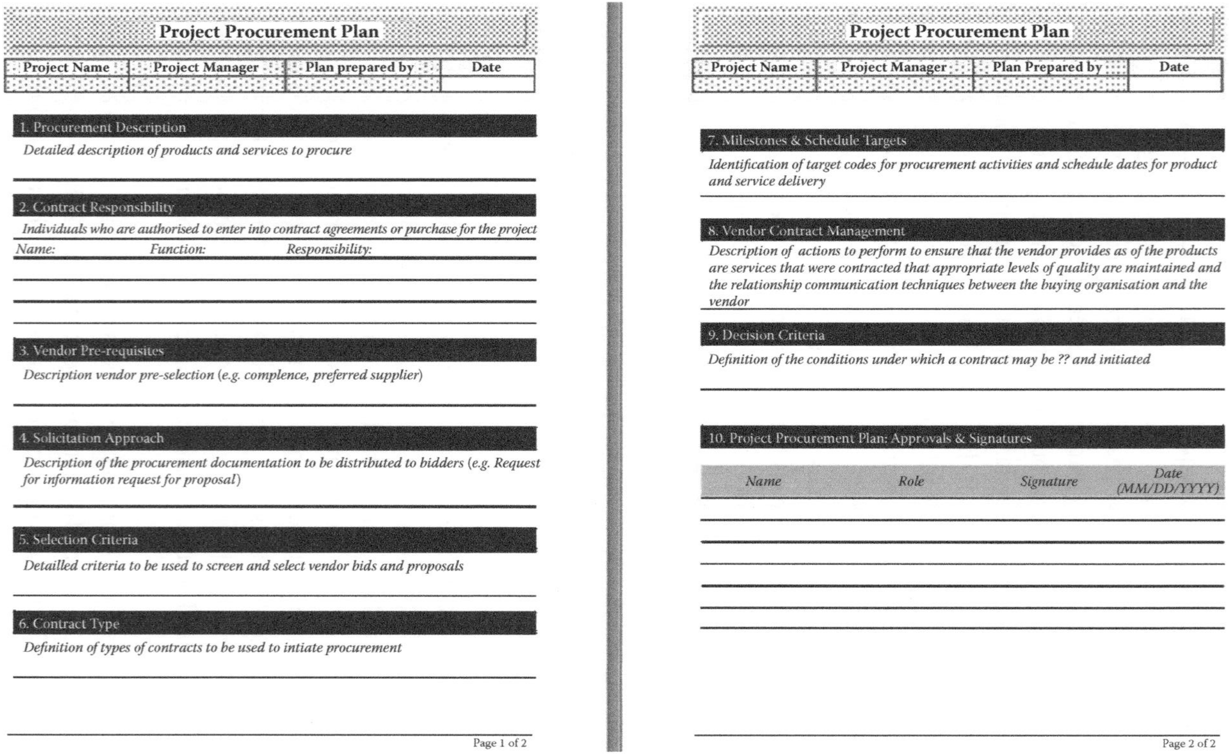

Project Procurement Plan

Project Name	Project Manager	Plan Prepared by	Date

7. Milestones & Schedule Targets
Identification of target codes for procurement activities and schedule dates for product and service delivery

8. Vendor Contract Management
Description of actions to perform to ensure that the vendor provides as of the products are services that were contracted that appropriate levels of quality are maintained and the relationship communication techniques between the buying organisation and the vendor

9. Decision Criteria
Definition of the conditions under which a contract may be ?? and initiated

10. Project Procurement Plan: Approvals & Signatures

Name	Role	Signature	Date (MM/DD/YYYY)

Page 2 of 2

Project Procurement Plan

Project Name	Project Manager	Plan prepared by	Date

1. Procurement Description
Detailed description of products and services to procure

2. Contract Responsibility
Individuals who are authorised to enter into contract agreements or purchase for the project
Name: Function: Responsibility:

3. Vendor Pre-requisites
Description vendor pre-selection (e.g. complence, preferred supplier)

4. Solicitation Approach
Description of the procurement documentation to be distributed to bidders (e.g. Request for information request for proposal)

5. Selection Criteria
Detailled criteria to be used to screen and select vendor bids and proposals

6. Contract Type
Definition of types of contracts to be used to initiate procurement

Page 1 of 2

FIGURE 6.2
Project procurement plan template

The plan should consist of the following elements (more can be added according to needs):

- Summary of the list of products and services to be procured based on the project procurement plan, specifications, quantities required, and cost estimated to be approved for procurement
- Identification of the procurement methods based on internal cost estimates where appropriate
- Establishment of a bidding committee, if needed
- Preparation of documents for the requisition step
- Issuance of invitation to tender and advertisement
- Opening of the bid process
- Evaluation of received bids and proposal for award of contract
- Award of contract
- Notification of award of contract
- Beginning and completion of negotiations and signing of contract
- Contract implementation
- Date, time of contract completion, or delivery

6.3.2 The Project Procurement Management Process

The project procurement management process consists of five key steps. These are explained in Figure 6.3 and detailed in the subsequent sections of this chapter.

FIGURE 6.3
Project procurement process

6.3.2.1 Requirement Step

A "make versus buy" analysis is conducted, a decision is made on the products or services to be purchased, the project procurement plan is prepared, and approval is given to proceed to the next step.

6.3.2.2 Requisition Step

Specifications for the product or service are developed, a statement of work (SOW) is established and a request for proposal (RFP) is completed. A decision is also made on the type of contract.

6.3.2.3 Solicitation Step

The SOW and/or the RFP are distributed and made available to the market and prospective providers. The received proposals are evaluated to select the most qualified vendor, according to established selection criteria. Negotiation with vendors concludes with the award of the contract.

6.3.2.4 Contract Administration Step

The contract and vendor performance are managed by working closely with the provider to ensure that contract requirements are met.

6.3.2.5 Closeout Step

The step terminates the procurement plan for the product or service. Verification is performed that the purchased items have been received and is acceptable, the status of any outstanding invoices is established and the final payment is made. Additionally, vendor performance is documented and lessons learned are documented. The contract is closed when the project procurement plan has been completely executed.

The detailed processes of the five key steps are documented in the project procurement plan. The plan must be monitored and updated to accommodate any changes enforced by changes in the marketplace and any changes to internal policy.

When product lead-times are long and materials are difficult to secure and deliver, the project procurement process will play a pivotal role. By implementing project procurement planning, the project manager will ensure that the company buys appropriate products and services at the most advantageous price from qualified suppliers to fulfill the project objectives. The project procurement process becomes more significant if the corporation's production operation is dependent on external vendors and suppliers to produce its own products and services.

6.3.3 Exercise: Describe Your project's Procurement Process

Review your current project and describe the procurement management process and plans in use.

6.4 DIFFERENT TYPES OF CONTRACTS

There are various types of contracts for different types of procurement. The type of contract used and specific contract terms and conditions set the degree of risk being assumed by both the buyer and seller.

Any contract will bind the product/service provider to supply/perform the defined statement of work, fulfilling the buying organization's project objectives.

The type of contract for the product/service to be purchased is identified by the buying organization during the requisition step of the project procurement process. This may eventually be modified during negotiation with the provider prior to final contract award. The project manager must ensure that all contracts protect the interests of both the organization and the project.

Contracts generally fall into one of four broad categories (see Figure 6.4):

- Fixed-price or lump-sum contracts
- Cost-reimbursable contracts
- Time and material (T&M) contracts
- Indefinite delivery contracts

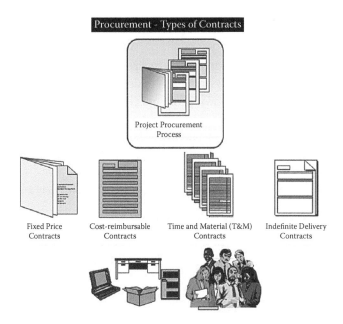

FIGURE 6.4
Procurement—types of contracts

6.4.1 Fixed Price or Lump-Sum Contracts

Outsourcing and turnkey procurements are generally done on a fixed price for deliverables basis (see Figure 6.5).

This type of contract is best used when the specific product and/or service to be delivered can be fully defined and specified before the start of work. The specifications are described in detail in a SOW, ensuring complete understanding of the requirements by both parties. The provider, or contractor, is required to successfully perform the specified work and deliver agreed-upon products and/or services within the prescribed price to the organization and agreed schedule and quality. The responsibility and risk for the delivery of the specific product/service is on the provider.

FIGURE 6.5
Contracts—fixed price

Fixed Price
Contracts

If the provider exceeds its internal contract cost, delivery of the product/ service to the outsourcing organization is still at the agreed-upon price. If the provider delays the agreed delivery date, costs to re-establish the original schedule and penalties incurred are at the provider's charge.

The project manager holds the responsibility for the detailed specification and the management of scope change. If the scope of the project is changed, a change request must be processed and the cost of the changes, which may not exceed established limits, must be agreed upon (see Chapter 7, Section 7.3, Scope Management and Change Request Management). When the project's statement of work and scope can be adequately defined, this type of contract is very favorable to the outsourcing organization for controlling its own costs, but changes in scope must be managed attentively.

This type of contract can include incentives for meeting or exceeding selected project objectives.

The simplest form of a fixed price contract is a purchase order for a specified item to be delivered by a specified date for a specified price.

6.4.2 Cost-Reimbursable Contracts

In some cases, usually when there are such uncertainties of performance that a price cannot be estimated with sufficient accuracy to ensure that it is fair and reasonable, cost reimbursement contracts may be preferred (see Figure 6.6).

Cost reimbursement contracts place the least cost and performance risk on the seller. They can regrettably only require the provider to use "best efforts" to complete the contract. These contracts are not often in the best interest of the buying organization, but may be useful in certain circumstances.

Cost-reimbursable
Contracts

FIGURE 6.6
Contracts—cost reimbursable

The advantages of the cost reimbursement contract are better control of project cost while still providing some flexibility when scope has not been fully defined.

Cost-reimbursable contracts involve payment, by the buying organization to the seller, of the provider's actual costs plus a fee typically representing a previously agreed seller's profit. This type of contract allows for payment to the provider of all costs incurred, within a predetermined ceiling and allowable cost standards, after the work of the contract is performed.

The contract cost is the aggregate of the seller's direct and indirect costs. Direct costs are costs incurred by the provider exclusively for the project execution. Indirect costs are usually calculated as a percentage of direct costs. Cost-reimbursable contracts often include incentive clauses for bonus payments. These are paid by the buying organization when the seller meets or exceeds selected agreed-upon project objectives, such as schedule targets or total cost.

The common types of cost-reimbursable contracts are cost-plus-fee (CPF), cost-plus-percentage-of-cost (CPPC), and cost-plus-incentive-fee (CPIF).

6.4.2.1 Cost-Plus-Fee (CPF)

With this contract, the buying organization pays the agreed cost and additionally a predetermined fee established at contract award.

Sellers receive a fee that varies with the actual cost, calculated as an agreed-upon percentage of the costs and also reimbursed for allowable costs for performing the contract work.

6.4.2.2 Cost-Plus-Percentage-of-Cost (CPPC)

This is somewhat similar to the CPF; however, the amount of the fee rises as the provider's costs rise.

This contract type provides no incentive for the seller to control costs, thus buyers rarely use it. Many government or public contracts specifically prohibit the use of this type of contracting.

6.4.2.3 Cost-Plus-Incentive-Fee (CPIF)

This type of contract is an extension of CPF, in which a larger fee is awarded to providers that meet or exceed performance targets including cost savings.

Payment of incentives, and disincentives, can be included in these types of contracts. They are based on predetermined performance standards that are agreed to by all parties to the contract. For example, an incentive fee could be included in a contract and awarded to a contractor if the product is delivered ahead of schedule, and a disincentive fee could be assessed if the product is delivered late.

CPIF contracts can also include cost savings sharing between the buyer and seller, if the final provider costs are less than the expected costs. The sharing applies a prenegotiated formula written in the contract.

To ensure that these items have a positive effect on the contract and to create a win-win situation, it is recommended that incentives only be used when the contract also includes disincentive clauses.

6.4.2.4 Time-and-Material (T&M) Contracts

Personnel services and product/materials procurements are generally made using a time-and-materials (T&M) contract (see Figure 6.7).

These contracts pay for services rendered at a fixed rate and for product/ materials at cost plus a handling fee. These types of contracts resemble cost-reimbursable-type arrangements in that they are open-ended. Buyer does not define the full value of the agreement and the exact quantity of items to be delivered at the time of the contract award. Thus, time and material contracts can grow in contract value as if they were cost-reimbursable-type arrangements.

This type of contract is usually used if the scope of the work to be completed is not well defined and does not permit a fixed level of effort or a fixed price to be estimated.

A contract is developed to secure services for a range of technical skills, with negotiated hourly rates. The contract is usually assigned a maximum amount payable. This contract type is particularly well suited to situations in which the principal "deliverable" is labor hours.

Time and Material (T&M)
Contracts

FIGURE 6.7
Contracts—time and materials

Contracted products or materials follow the technique used for indefinite delivery contracts (Section 6.4.3). The main difference lies in that T&M contracts define a maximum volume not to be exceeded for the project.

The project manager must provide for the management of each individual contract staff person's performance, and specific performance standards for each type of resource must be established. This may be particularly difficult if different performance standards are being used for contract staff and for in-house staff performing the same functions.

In this type of contract, the burden is on the project manager to control project scope and cost by defining individual performance standards and monitoring contractor performance.

6.4.3 Indefinite Delivery Contracts

Indefinite delivery contracts provide for delivery of products and services, upon the issuance of a delivery order when specific needs arise (see Figure 6.8).

Many corporations use these predefined contracts to simplify the procurement process. These contracts are sometimes known as "on call" or "term."

Utilizing an indefinite delivery contract minimizes the requirement to establish terms and conditions per purchase, since it is only necessary to go through the process of defining the requirements once at contract award. In many cases, the buying organization will enter into a contract before the project starts.

These contracts usually state the type of product/service to be provided, a length of time that the service can be requested, and a maximum contract amount. The contracts can significantly reduce the amount of time it will take to secure products/services, but are normally only used on smaller, less complex projects.

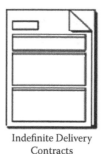

FIGURE 6.8
Contracts—indefinite delivery

Indefinite Delivery
Contracts

6.4.4 Exercise: Types of Contracts on Your Project

Review your current project and describe the types of contracts that have been awarded for the project's products and services.

6.5 PROCUREMENT PROCESS: REQUIREMENT

The requirement step identifies which project needs can best be met by purchasing products or services from outside the project organization (see Figure 6.9). A "make or buy" analysis is usually conducted to establish which products and services need to be purchased. This guides what to procure, how to proceed, the quantities or volumes to purchase and when to acquire these. This information constitutes the body of the project procurement plan, which drives the total project procurement process from requirement through to contract closeout for each product/service item to be purchased or acquired.

During the requirement step, potential providers are identified who are aligned to the organizational policy and are compliant with current legislation and regulations.

FIGURE 6.9

Process step—requirement

The requirement step includes reviewing the risks involved in each purchasing decision and leads to selecting the type of contract planned to be used with respect to mitigating risks and transferring risks to the provider.

6.5.1 Detailed Requirement Step Process

- Decide which products and services will be procured and under what conditions. Determine if these are present in the organization, and substantiate why they will not be able to support project needs.
- When necessary, create a project procurement team (program manager, project manager, project subject matter experts, procurement personnel, IT experts, business partners, legal experts, finance, HR, etc.).
- Decide on who, from the project team and organization unit, can interface with the vendors and can sign the contract. Adherence to organization procurement policies have to be considered.
- Establish prequalification evaluation criteria for vendors.
- Identify the presence of "preferred vendor lists" limiting the number of vendors solicited. Similarly, "minority groups" or other legal and regulatory obligations must be considered, especially for open public contracts.
- Identify the procurement method(s), i.e., lease/purchase, bid process, etc.
- Identify the individuals who must approve any purchases.
- Provide target dates for all the relevant procurement activities, to ensure that vendors have resources available to meet the project schedule.
- List the required capabilities of the product or service. These describe the capacity and/or volumes to be handled.
- Identify the documentation, manuals, and training that will be necessary for delivery and operation of the product or service.
- Describe the vendor's method of post-delivery supportability, warrantees, and guarantees needed for operational use of the product or service (see Figure 6.10).

6.5.2 Make-or-Buy Analysis

The make-or-buy analysis results in the choice whether a particular product or service can be produced by the project team internally (in-house) or can be procured externally (from an outside supplier). The decision to buy is referred to as outsourcing.

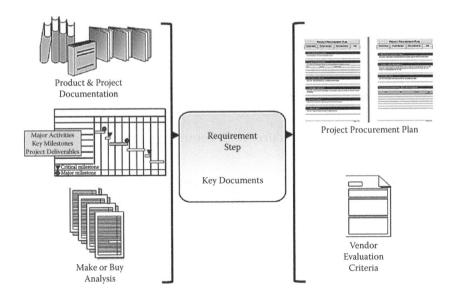

FIGURE 6.10
Key documents—requirement

Make-or-buy decisions usually arise for a project when an organization requires a product or service that it may not be able to provide, is too expensive to perform in-house, or is not suitable to make as not compatible with the project's goals.

Factors that may influence organizations to procure include:

- Cost considerations (less expensive to buy the item)
- Lack of expertise
- Suppliers' specialized know-how exceeds that of the buyer
- Small-volume requirements
- Desire to maintain a multiple-source policy
- Indirect managerial control considerations
- Item not essential to the organization's strategy

If a buy decision is to be made, then a further decision of whether to purchase or rent is also made. The analysis includes both direct as well as indirect costs.

The performing organization may reach a buy decision to procure a product or service that may not be cost effective from the perspective of the project but has validity for the organization. For example, purchasing an item for the company's use (anything from a piece of heavy equipment to contracting external staff), rather than renting or leasing it for the sole use of the project. In this case, the cost allocation to the project would be based upon a margin analysis.

6.5.3 Documenting Procurement Requirements: Statement of Work

A statement of work (SOW) is defined for each product or service to be procured, and refers to the corresponding project scope specification and the related project work breakdown structure. The SOW is to be written clearly, completely, and concisely, and describes the products or services to be provided by the vendor. This should give sufficient detail to allow prospective sellers to determine if they are capable of providing the product or service: specifications, quantity desired, quality levels, performance data, period of performance, work location, and other requirements. Any concomitant services required, such as performance reporting or post-project operational support is to be included in the SOW.

The statement of work may be revised and refined as required, either because of greater detail and knowledge from the project planning activities or from prospective seller suggestions.

6.5.4 Specific Guidelines for Product Procurement

In the case of purchase of products, details of the requirement should be described using generic, comprehensive, and unambiguous technical specifications. Specifications must be clear and sufficiently detailed to enable vendors to meet the identified need and to compete fairly. Except in the case of legislation, specifications should not refer to brand names, catalogue numbers, or types of equipment from a particular manufacturer. When it is necessary to specify a particular essential design, or characteristic of functioning, construction, or fabrication, the references should be followed by the words "or equivalent" together with the criteria for determining such equivalence. The specifications should permit the acceptance of offers for equipment with similar characteristics that provide performance and service at least equal to that specified.

6.5.5 Vendor Evaluation Criteria

Vendor evaluation is best performed in at least a two-phase process. The first phase is a prequalification phase and is performed during the requirement step. The second phase is performed during the requisition and solicitation steps (see Sections 6.6 and 6.7).

Potential bidders register an "expression of interest," request and submit a prequalification questionnaire (PQQ). Suppliers that meet the PQQ criteria receive an invitation to tender (ITT) at the requisition step.

6.5.5.1 Prequalification Questionnaire

The prequalification questionnaire (PQQ) contains further information about the procurement, together with a range of questions and requests for information, where financial and nonfinancial factors are considered.

The criteria are:

- Supplier acceptability: For instance, whether a company is in receivership or bankrupt, and whether the organization or directors have been convicted of a criminal offense relating to the conduct of the business.
- Economic and financial standing: Financial checks to ensure that suppliers are in a sound financial position to participate in the intended procurement.
- Technical capacity: Suppliers must show a track record demonstrating a standard of technical capacity relevant to the procurement.

Suppliers, and previously qualified sellers, submit their PQQs. Each is evaluated by the evaluation team, and the scores are moderated to arrive at an overall mark. Those that pass the criteria proceed to the requisition step.

Suppliers that do not proceed beyond the PQQ stage are notified in writing and offered a debrief meeting to identify areas in which their submission did not meet the evaluation criteria.

Note that it is common for organizations to maintain a preferred vendors list that includes only sellers already selected from previous procurement cycles. These preferred vendors may also have already established blanket corporate agreements that cover both the company's global and local procurement needs.

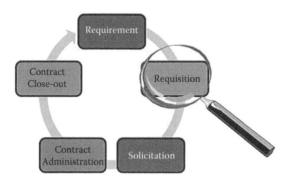

FIGURE 6.11
Process Step - Requisition

6.6 PROCUREMENT PROCESS: REQUISITION

The requisition step (Figure 6.11) concerns the preparation of the documents needed to support the next solicitation step. The project manager and the procurement team develop full specifications for the product or service to procure; the corresponding SOW are completed, and the qualified supplier evaluation criteria checklist is finalized. All these documents are collated to form a solicitation package.

A preliminary schedule is drafted for meetings and conferences with prospective sellers. The solution package is verified to conform to the current project scope and plan and requisitions orders are prepared. All requisitions will require approval before the solicitation packages are made available to qualified potential suppliers.

The solicitation package is prepared during this requisition step but utilized during the solicitation step. In all situations, the same solicitation package must be sent to each possible supplier for a particular product/service, so that the playing field is level.

In this step, a solicitation strategy must be coherent with the type of solicitation package to develop. Depending on the nature of the project and the product/service to procure, the project manager can select amongst the major following schemes:

- **Request for expression of interest:** Also known as a "registration of interest," the request of interest serves to determine the level of interest in the business community and to obtain comments from enterprises about the future project. This step is optional.

- **Request for qualification:** The candidates that participate in the request for qualification must demonstrate their technical capacity to provide and/or perform the product/service; their organizational and technical capacity to operate and maintain the product/service in the buyer's operational context, their capacity to fund the work and their compliance to legal, statutory, and regulatory regulations.
- **Request for proposals or request for quotations:** The requests allow prospective vendors and suppliers that meet the requirements of the request to present specific tenders or proposals that must contain a technical component and costs.

A listing of qualified vendors that are requested to bid is included in order to drive down the cost where possible. Quite often, one vendor may not bid, because it recognizes that it cannot submit a lower bid than one of the other vendors. The cost of bidding is an expensive process for the vendor.

A schedule of bidders' conferences is used so that no single bidder has more knowledge than others. For most procurement, the conferences can be informal meetings to clarify the contents of the solicitation package. For government contracting or public tenders, if a potential bidder has a question concerning the solicitation package, then it must wait for the bidders' conference to ask the question so that all bidders will be privileged to the same information. There may be several bidders' conferences between solicitation and award.

Project managers may or may not be involved in the bidders' conferences, either from the procurer's side or the supplier's side.

A typical solicitation package would include:

- Bid documents (usually standardized)
- Vendor proposal evaluation criteria
- Bidder conferences schedule
- Listing of qualified vendors (expected to bid)
- How change requests will be managed
- Supplier payment plan

6.6.1 Detailed Requisition Step Process

- Verify that necessary product/service specifications have been provided.
- Prepare solicitation package with appropriate members of the project procurement team.

- Utilize standard solicitation documents (consistent with corporate policies) to develop:
  - Request for Information (RFI)
  - Invitation to tender (ITT)
  - Request for quotation (RFQ)
  - Request for proposal (RFP)
    - The RFP is the most costly endeavor for the vendor. Large proposals contain separate volumes for cost, technical performance, management history, quality, facilities, subcontractor management, and others.
- Establish vendor evaluation criteria, with clear, concise definitions for each criterion to facilitate common understanding.
- Develop a detailed, mathematically sound scoring plan that explains how proposals will be evaluated, and provide specific description of the scoring methodology. (For example, if a certain criterion is worth *x* points, a description should follow of what the points signify.)
- Determine if a presolicitation conference will be used to screen potential vendors.
- Draft a bidders' conference schedule, indicating calendar dates, times, and formality of meetings.
- Develop a change request management plan.
- Develop a vendor payment plan.
- Raise requisition orders and review these by the procurement team. Reach final approval from project sponsor and procurement department.
- Provide the solicitation packages to sponsor and key stakeholders for review and approval. (See Figure 6.12.)

6.6.2 Establishing Vendor Evaluation Criteria

A vendor evaluation criteria checklist represents how potential suppliers will be selected for the product/service to be procured. This checklist can be a standard form to cover all procurement items or can be specifically tailored to meet particular demands.

Each criterion answer can be simply stated as yes/no or can be weighed or scaled. It is imperative that the vendor evaluation technique be included in the solicitation package and known to prospective sellers, to ensure fairness and to limit claims.

Many techniques can be used to build the evaluation criteria. A simple technique is to establish a list of criteria against which a yes/no can be

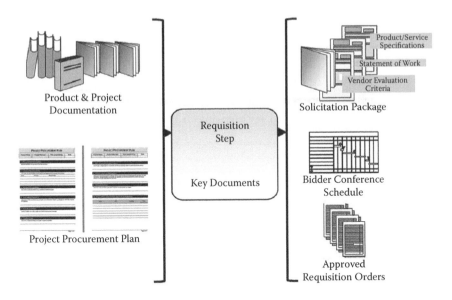

FIGURE 6.12
Key documents - Requisition

scored. A more advanced common technique is to determine major categories and breaking down each to identify detailed criterion elements. Then each criterion can be scored against a scale and comparison across vendors can be performed.

6.6.3 Examples of Vendor Evaluation Criteria Questions

The examples in this section illustrate how the project manager and the procurement team can prepare a coherent and appropriate list of criteria questions. It is to be noted that each evaluation criteria would need to be adapted to the particular product or service to procure, as well as the size and type of project concerned.

Adapted from PMBOK:

- Does the proposal address the contract statement of work?
- Do the proposed technical methodologies, techniques, solutions, and services meet the procurement documentation requirements?
- Do references exist from previous customers, verifying the seller's work experience and compliance with contractual requirements?
- Do the necessary financial resources exist?
- Do project management processes and procedures exist?

- Is the proposed total life cycle cost (ownership and operational cost) the lowest?
- Are technical skills and knowledge adequate for the job?
- Can potential future requirements be met?
- Do intellectual property rights exist on the products or services proposed?
- Is there compliance to regulations to meet a specific type or size of business?

Major Categories (Example 1)

MAJOR EVALUATION CRITERIA

Business reputation and references
Cost management history
Cultural congruence to buyer's organization
Customer relationship and satisfaction record
Delivery history for products & services
Desire for business
Domestic socioeconomic and political stability
Environmental and social responsibility
Financial status
Geographical coverage
Health and safety awareness
Information technology and communication systems
Innovation and R&D capability
Labor relations history
Management styles and organization
Market position and share
Performance history
Personnel development schemes
Procedural and regulatory compliance
Product reliability
Production facility and capacity
Project management experience
Quality systems
Strategic fit to buyer's organization
Technical capacity and support services
Warranties and claim policies

Major Categories—Breakdown (Example 2)

- Compliance to specifications
  - Meets specification requirements
  - Meets standards
- Compliance to contract terms
  - Complies to the terms and conditions
  - Existing or potential conflict of interest
  - Legal proceedings related to contractual issues—past or present
- Total cost of ownership
  - Bid price
  - Price breaks and quantity discounts
  - Warranty
  - Operational and maintenance costs (applying net present value analysis)
- Past performance
  - Experience in the industry
  - Customer recommendation
- Capability
  - Staffing structure
  - Availability of experienced staff
  - Experience in the industry
  - State of technology
- Financial review
  - Satisfies key financial ratios for the industry
  - Accounting principles
  - Lease versus buy
  - Foreign exchange
  - Payment terms and methods (i.e., EDI, etc.)
  - Insurance
- Risk and insurance
  - Adequate insurance
  - Allocate and acceptance of risk
- Customer service
  - Policy and practice
  - Customer surveys
  - Systems to measure customer satisfaction

- Quality system for deliverables
- Certification
- Documented system
- Geographical
  - Location
  - Networking
- Innovation
  - Leading technology
- Creativity

6.6.4 Scoring/Weighting Vendor Evaluation Criteria

The scoring/weighting scheme to use during the solicitation step is included in the solicitation package forwarded to prospective sellers. As was stated previously, the scoring could be a yes/no answer or a scaled score with or without a weighting, as illustrated in the next two sections.

6.6.4.1 Simple Scaled Score

Each criterion is scaled for all supplier responses, using a scale similar to the one presented in Figure 6.13. Ranking is then performed on the aggregate total for each supplier (see Figure 6.14).

6.6.4.2 Weighted/Scaled Score

The technique is as the simple scaled score; however, the project manager and the procurement team decide on which major category or criterion to apply a specific weight, which highlights its importance in the selection process (see Figure 6.15).

Criteria	Score	Average Weighted Scores			
	0 to 5	Supplier 1	Supplier 2	Supplier 3	Supplier 4
		0	0	0	0
		0	0	0	0
		0	0	0	0
		0	0	0	0
		0	0	0	0
		0	0	0	0
Total per Supplier		0	0	0	0

FIGURE 6.13
Vendor evaluation

Score	Scoring Scale & Description
0	Capability not demonstrated
1	Limited demonstration of capability & significant gaps identified
2	Partial demonstration of capability but gaps remaining
3	Capability demonstrated with no gaps
4	Capability clearly demonstrated with some value added benefits
5	Superior demonstration of capability with significant added benefits

FIGURE 6.14
Vendor evaluation weights

6.7 PROCUREMENT PROCESS: SOLICITATION

The solicitation step (see Figure 6.16) is central to the whole project procurement process. It is during this step that the solicitation package is issued to the prequalified and/or prospective vendors, who respond with their proposals, tenders, and bids. On receipt of all proposals, the buying organization proceeds with vendor evaluation, replying to questions and clarifications when necessary, selecting the best-value proposal and negotiating with the most qualified vendor to finalize the procurement with a contract award.

Competitive procurements utilize either a request for quotation (RFQ) or a request for proposal (RFP) process. Generally, product commodities are awarded on the basis of lowest price, as a result of an RFQ. While services or technology may be awarded on lowest price as well, they are more often awarded on the basis of best value as a result of an RFP process.

Criteria	Max. Score	Criteria Weighting	Supplier 1			Supplier 2			
			Score	Avg. Score	Weighted Avg.	Score	Avg. Score	Weighted Avg.	
	5	1.00	5.00	5.00	5.00	5.00	5.00	5.00	5.00
	5	1.00	5.00	5.00	5.00	5.00	5.00	5.00	5.00
	5	1.00	5.00	5.00	5.00	5.00	5.00	5.00	5.00
	5	1.00	5.00	5.00	5.00	5.00	5.00	5.00	5.00
	5	1.00	5.00	5.00	5.00	5.00	5.00	5.00	5.00
	5	1.00	5.00	5.00	5.00	5.00	5.00	5.00	5.00
Totals				30.00	30.00		30.00	30.00	

FIGURE 6.15
Supplier evaluation spreadsheet

FIGURE 6.16
Process Step - Solicitation

The project manager should consult with procurement staff prior to initiating a competitive procurement. This may able to save significant time and effort for the project team by collecting historical data from other projects that have already developed similar RFPs or RFQs. These previous projects can provide valuable information on evaluation and selection methods and processes, good contractual language, and performance standards. By utilizing established procurement best practices, the project manager can more effectively and efficiently complete the procurement.

6.7.1 Detailed Solicitation Step Process

- Issue the solicitation package
- Receive and answer questions from vendors:
  - In a public forum for bidders conferences
  - In formal or informal meetings
- If necessary, prepare for vendor product demonstrations, or oral presentations
- Receive and evaluate proposals
  - Evaluate proposals using predefined evaluation criteria
  - Establish in-range proposals
    - Inform in-range vendors of preselection

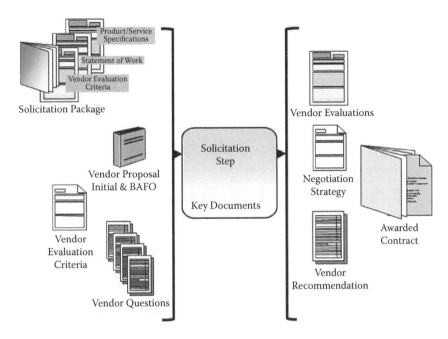

FIGURE 6.17
Key documents - Solicitation

- Inform vendors that are out of range (who may present a second proposal)
- Before "best and final offer" (BAFO) negotiation with in-range vendors:
  - Gather and assess all the facts on the proposals
  - Perform a complete cost analysis
  - Evaluate each vendors competitive position
  - Define your negotiation strategy and tactics
    - Develop minimum-maximum positions
    - Seller minimum willingness to accept
    - Buyer maximum willingness to pay
    - Establish negotiation setting
      - Office or other location
      - Morning or afternoon
      - Equipment and visual aids
      - Attendance and seating arrangements
      - Order of the vendor meetings
- Request BAFO
- Reevaluate offers using scoring method to select best value

- Review/redefine payment schedule
- Make vendor selection recommendation to sponsor and procurement management to reach agreement and approval
- Sponsor and procurement management review recommendation and approve contract award
- Project manager awards and signs contract, then issues purchase order

6.7.2 Issue of Solicitation Package

The solicitation package is the key document issued by the buying organization; its primary content covers the SOW and the specifications of the product/service to procure.

The buying organization issues the solicitation package to an established set of prequalified vendors, as identified during the requisition step. The package must be comprehensive and clearly described.

Specific vendor selection criteria, the evaluation instruments, the weighting of individual elements, the cost component, and the technical component must all be defined before issuance of the package and receipt of proposals. It is paramount that this process provides equal and fair opportunity to each proposal.

The solicitation package should contain a clearly marked section that describes the invariable and mandatory terms and conditions of the contract. This will include such items as the method of payment, required project schedule, location of work, method of product delivery, warranties, and damages for nonperformance, etc. Furthermore, the solicitation package must clearly explain the administrative aspects of the procurement, such as contacts, key dates, policy and bidding requirements, format of proposals, etc.

6.7.2.1 Acquisition Method

Selection of the acquisition method is the critical element in the solicitation step. There are two common methods for acquisition:

- **Competitive** is when market forces determine the price and the award goes to the lowest bidder. There are no negotiations. Used often for sealed bids.
- **Negotiation** is when the price is determined through a bargaining process.

6.7.2.2 Advertising

Competitive acquisition is normally awarded following public adver-tisement as part of a formal tendering process. Publication on local and national advertising media is the most common instrument. The general use of Web sites now supersedes the standard media. Depending on the nature of the requirement, it may be appropriate to supplement national and Web site advertising with advertising in other media, such as in trade publications and/or other related Web sites.

- **Open** is where any vendor can request a copy of the solicitation package. There is no control over the number of tenders requested or submitted.
- **Restricted** is used to restrict the number of vendors invited to respond to the solicitation. A Prequalification Questionnaire (see Section 6.5.5.1, Prequalification Questionnaire) is issued to all companies that have expressed an interest. The responses are then evaluated against pre-set criteria with invitations to respond issued only to those companies identified as most suitable to the requirement. The maximum number of bidders must be declared at the outset on the advertising.

6.7.2.3 Bidders' Conference

Bidders' conferences are held with prospective vendors. They are used to ensure that all bidders have a clear, common understanding of the pro-curement and that potential sellers are given equal standing during this initial buyer and seller interaction to produce the best bid.

The bidders' conference is hosted by the buying organization, and every selected vendor receives a request for response (RFR). Prospective bidders, who accept attendance by completing a bidders' conference registration form, are typically welcome to submit written questions. The RFR will indicate where and when a bidders' conference will be held and if writ-ten questions will be accepted. The answers for all questions are generally posted prior to the conference. The date for a bidders' conference and the deadline for written inquiries are indicated in the procurement calendar included in the RFR.

Potential sellers who have completed a bidders' conference registration form are expected to attend. On many occasions, attendance at the bid-ders' conference is a mandatory requirement to enable the vendor to sub-mit a bid.

At the bidders' conference, the buying organization introduces key individuals in the procurement process, explains the procurement rules, presents the contents of the solicitation request, and answers procedural questions. Technical questions are not normally answered at the conference and must be written and submitted prior to the event.

6.7.3 Receipt of Vendor Proposals

The project manager and the procurement team should ensure that proper procedures are in place for opening tenders, which prevent any abuse at this stage. All tenders should be opened together as soon as possible after the designated latest time and date set for receipt of tenders. Internal procedures should require that opening of tenders takes place in the presence of at least two individuals. The procedure adopted should ensure that, in the case of any dispute, there is a clear and formal independently vouched report of the tenders received. Tenders received after the closing time for receipt of tenders should not be accepted.

Prior to receipt of vendor proposals, an independent estimate of the time and cost to complete the project should be developed. This should be a realistic and not overly optimistic estimate that takes into consideration the technologies and skills involved in the project, especially if new technologies are involved. This is occasionally done through an RFI process prior to the decision to develop the procurement solicitation, to establish a baseline for cost comparison. It will also assist to assess the available competitive field and state of the marketplace for the particular industry.

This independent estimate will provide a baseline for comparing proposals during vendor selection. If there are significant variations between cost and schedule estimates in submitted proposals, the lowest bid may not always be the best value. If the low bid is significantly lower than the independent estimate, it should be looked at very carefully. If the contract is not profitable it can generate many problems for the vendor and the buying organization.

6.7.3.1 Answering Questions

Vendors may need to clarify certain areas of the procurement package. This should not be allowed in the case of sealed tenders or after a bidders' conference.

However if the acquisition method is through negotiation, then the buying organization can schedule formal or informal meetings with the vendor to elucidate or explain parts of the RFP. This may at times also allow the vendor to discuss different proposal options that can be considered.

6.7.4 Vendor Proposal Evaluations

Vendor proposals are evaluated and preferred contractors or suppliers are selected, in accordance with previously set and agreed selection processes and organizational requirements.

Generally, there are three steps in the evaluation and selection process:

1. A technical evaluation review to evaluate the technical and functional aspects of the proposals
2. A cost or financial evaluation review to evaluate the total acquisition cost proposed
3. A recommendation report of both evaluations forwarded to the sponsor and/or procurement management to authorize the final decision and proceed to contract award.

Ideally, the contract language included in the RFP will be of sufficient detail that limited negotiation is required after a proposal has been selected.

Each received vendor proposal is evaluated using the pre-defined evaluation criteria (see Section 6.6.2, Establishing Vendor Evaluation Criteria).

6.7.5 Vendor Selection

Vendors that pass the evaluation criteria are categorized in the "in-range" proposal list, and the vendor providing best-value from this list is selected. "Out-of-range" vendors are eliminated.

6.7.5.1 Inform Vendors of Selection

"In-range" selected vendors are informed by formal means and invited to negotiate their proposals.

"Out-of-range" vendors are also informed that they are no longer considered in the procurement selection process. The buying organization may decide, if and when these vendors fail to pass selection by a close margin, but whose proposal presents certain key advantages, to submit a second proposal.

6.7.5.2 Prepare Negotiation with "In-Range" Vendors

Negotiations with selected in-range vendors and suppliers should be planned. A typical list of activities would include:

- Be conversant on the details of the vendor's proposal and terms and conditions
- Develop objectives of the negotiation and establish minimum and maximum positions
  - From the buyer perspective: what is the maximum to pay?
  - From the seller perspective: what is the minimum to accept?
  - Evaluate the vendor's strengths and weaknesses
    - This can be done by using a SWOT analysis, which also high-lights threats and opportunities
    - The project manager should determine what motivates the vendor: interest in profitability, keeping staff employed, developing a new technology, using the buying organization's name as a reference, entry into the market, etc.

- Perform a complete price/cost analysis, including cost of acquisition, and where necessary, the cost of ownership in operations
- Define the negotiating strategy and tactics
- Establish negotiation setting
  - Office or other location
  - Morning or afternoon
  - Equipment and visual aids
  - Attendance and seating arrangements
- Order of the vendor meetings

6.7.6 Negotiation with Vendors

Negotiations are conducted with the selected vendors, contractors, or suppliers. The negotiators can request guidance of senior personnel, if required, to agree on contract terms and conditions, establish common goals, and minimize uncertainty, in accordance with organizational requirements.

Negotiation ends when the buying organization selects the vendor that offers the best value for the requested procurement product/service. The project manager then submits the vendor selection recommendation to sponsor and procurement management for agreement and approval.

Vendor relations are critical during contract negotiations. The integrity of the relationship and previous history can shorten the negotiation process. The three major factors of negotiations are:

- Compromise ability
- Adaptability
- Good faith

The project manager should hold a post-negotiation critique in order to review what was learned. The first type of post-negotiation critique is internal to the buying organization. The second type of post-negotiation critique is optional and is conducted with all of the losing bidders to explain why they did not win the contract. Losing bidders may submit a "bid protest," in which the buying organization may have to prepare a detailed report as to why this bidder did not win the contract. Bid protests are most common on government and public contracts.

6.7.7 Contract Award

Contract award is often called the award cycle. The objective is to negotiate a contract type and price that will result in the performance and delivery of the product/service to the buying organization, whilst offering reasonable risk to the vendor and provide the vendor with the greatest incentive for efficient and economic performance.

The contract award results in a signed contract between both buying and selling parties. The type of contract indicated in the solicitation package serves as the basis. However, during the negotiation process the type of contract and/or contents may be modified. The final type of contract selected is based upon the following:

- Vendor's responsibility (and risk)
- Degree of cost and schedule risk
- Type and complexity of requirement—technical risk
- Extent of price competition
- Cost/price analysis

- Urgency of the requirements
- Performance period
- Vendor's accounting system
- Concurrent contracts—impact of other contracts currently performed by the vendor that can jeopardize performance on this contract
- Extent of subcontracting—proportion of this contract that the vendor will outsource

Whatever contract is awarded, the basic elements consist of:

- Mutual agreement: There must be an offer and acceptance
- Legal purpose: The contract must be for a legal purpose
- Form provided by law: The contract must reflect the contractor's legal obligation, or lack of obligation, to deliver end products
- Consideration: There must be a down payment
- Contract capability: The contract is binding only if the contractor has the capability to perform the work

The two most common forms of contracts are completion contracts and term contracts.

- **Completion contract:** The contractor is required to deliver a definitive end product or service. Upon delivery and formal acceptance by the customer, the contract is considered complete, and final payment can be made.
- **Term contract:** The contract is required to deliver a specific "level of effort," not an end product. The effort is expressed in effort hours (days, months, or years) over a specific period of time using specified personnel skill levels and facilities. When the contracted effort is performed, the contractor is under no further obligation. Final payment is made, irrespective of what is actually accomplished technically.

The final contract is usually referred to as a definitive contract, which follows normal contracting procedures such as the negotiation of all contractual terms, conditions, cost, and schedule prior to initiation of performance.

Regrettably, negotiating the contract and preparing it for approvals and signatures may require months of preparation. If the buying organization needs the vendor to begin immediately or if long-lead product procurement is necessary, then the buying organization may provide the contractor with a letter contract or letter of intent. The letter contract is a preliminary written instrument authorizing the contractor to begin immediately the supply or manufacture of products or the performance of services. The final contract price may be negotiated after performance begins, but the contractor may not exceed the "not to exceed" face value of the contract. The definitive contract must still be negotiated.

6.7.8 Exercise: Awarded Contracts on Your Project

Review your current project and list the products and services contracts that have been awarded for the project and the terms and conditions that have been agreed.

6.8 PROCUREMENT PROCESS: CONTRACT ADMINISTRATION

The purpose of the contract administration step is to ensure that the vendor (contractor) and the products or services delivered comply with the contract requirements (see Figure 6.18). Each project should be assigned a contract administrator, who may be the project manager or an assigned member of the project team.

The keys to successful vendor contract administration are an unambiguous and mutual understanding of the contract and a good business relationship.

As outlined in Section 6.3.1., the Project Procurement Management Plan is the key document that describes the method in which contracts will be administered and executed. The plan will above all explain the requirements of vendor performance.

The performance standards for the contractor must be articulated in the contract, and the vendor should demonstrate a complete understanding of the standards and show that a process has been established for meeting

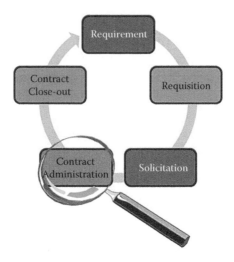

FIGURE 6.18
Process step—administration

each of these standards. There are benefits to both the vendor and the buying organization if a positive relationship can be established and maintained in which risks and benefits are shared.

After a vendor has been selected to provide products or services, the project manager is responsible for managing the contractor's performance, either directly or through a team member specifically assigned to the task. The project manager must ensure that responsibility is established for vendor oversight, and that the vendor receives clear direction regarding all contractual matters.

Progress of procurement is reviewed and agreed changes are managed to ensure timely completion of tasks, resolution of conflicts, and achievement of project objectives in accordance with the contract.

6.8.1 Detailed Contract Administration Step Process

- Interpret specifications
- Ensure that quality of the product or service is maintained
- Manage product warranties
- Manage subcontractors (if required by the contract)
- Direct change management
  - Administrative changes
  - Budgetary changes

- Contract modification (if necessary) and associated change order with proper approvals
  - Manage smaller, incidental contracts for work associated with the project
- Resolve contract disputes
  - Part of the work may be accepted
  - All of the work may be rejected
  - Work may be accepted with provisions for corrections in the future
- Completion of the project
  - Ensure that all project requirements are complete per contract
  - Product is technologically out of date (contract language should guarantee current technology)
- Terminate the contract
  - Default of contract
  - Contractor fails to perform any provision of the contract including:
    - Failure to deliver by scheduled date
    - Failure to maintain costs within a given threshold
    - Failure to make progress (endangers performance of the contract)
  - Termination for convenience (e.g., project cancelled)
  - Document contract problems and file formal vendor complaint with procurement management, e.g., liquidated damages, opportunity cost, etc. (if necessary)
- Manage contract risk associated with the project (see Figure 6.19)

Finalization activities are conducted to ensure contract deliverables meet contractual requirements, in accordance with project plan and organizational requirements.

Project deliverables are reviewed using available procurement records and information to determine the effectiveness of contracting and procurement processes and procedures, in accordance with organizational requirements. Procurement records and information may include product specifications; procurement management plan; contract documentation; contractor selection criteria, processes, and recommendations; contract negotiation documentation; contract change proposals and approvals; test and acceptance procedures and documentation; contract discharge and asset disposal register.

Any procurement or contract management issues and recommended improvements are identified, documented, and passed on to senior

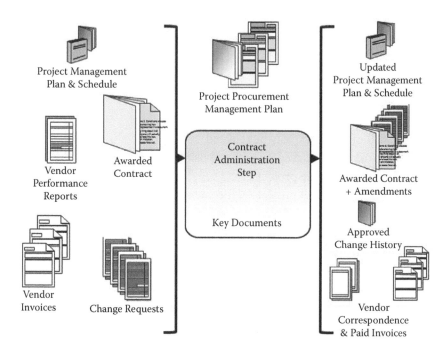

FIGURE 6.19
Key documents—contract administration

personnel for application in future projects in accordance with organizational requirements.

6.8.2 Managing the Relationship with the Vendor/Seller/Supplier

The project manager should hold regularly scheduled meetings with the vendor to obtain information about how effectively the contractor is achieving the contractual objectives. Periodic reviews with the contractor should be established to ensure contractor adherence to standards and compliance with project processes and schedules.

The project manager and team members should entertain a climate of open discussions and mutual trust and consideration with vendors.

6.8.3 Change Request Management

Changes will occur in all projects and these have to be managed diligently with the vendors. Change request management is discussed in Chapter 7,

Project Implementation. For the purpose of this chapter, it is imperative that the project manager maintains an up-to-date change request history log with each vendor and the corresponding contract. Changes to the scope of work, cost basis and/or scheduled delivery will have an impact to the current contractual terms. The project manager, or contract administrator, must measure these impacts with the vendor and how they align to the established performance criteria of the contract. Contracts would need to be modified to reflect the approved incorporated changes.

6.8.4 Contract Disputes

Even with the best commercial intentions and the most accurately drawn-up agreements, contract disputes are an inevitable part of project life. If handled poorly, they can result in a loss of customer/supplier goodwill, management time and money.

Disputes may arise over quality of work, over responsibility for delays, over appropriate payments due to changed conditions, or a multitude of other considerations. Resolution of contract disputes is an important task for project managers. The mechanism for contract dispute resolution can be specified in the original contract or, less desirably, decided when a dispute arises.

Negotiation among the buying and selling parties is a preferred and strongly suggested dispute resolution mechanism. The negotiation process is usually informal, unstructured, and relatively inexpensive. These negotiations can involve the same sorts of concerns and issues as with the original contracts. Negotiation typically does not involve third parties such as lawyers and judges. If an agreement is not reached between the parties, then both parties would pursue the dispute through legal representation.

While various dispute resolution mechanisms involve varying costs, it is important to note that the most important mechanism for reducing costs and problems in dispute resolution is the reasonableness of the initial contract among the parties as well as the competence of the project manager.

6.8.4.1 Disputes due to Interpretations

In general, most contract disputes occur as a result of disagreements over the meaning of terms in the contract. Disputed terms can involve design or performance specifications, the statement of work or even the schedule.

6.8.4.2 Disputes in Work Instructions

Disputes in work instructions often result when the buying organization considers that the work was not authorized either because the contractor misinterpreted the instructions or the individual who gave the instructions was not authorized to do so. Other reasons that occur for this type of dispute are due to the buying organization and the vendor frequently disagreeing as to whether the work entitles the contractor to a reasonable adjustment of the price and/or the schedule.

6.8.4.3 Adjudication

Adjudication is the legal process by which an arbiter or judge reviews evidence and argumentation including legal reasoning set forth by opposing parties or litigants to come to a decision which determines rights and obligations between the parties involved.

This process tends to be expensive and time consuming since it involves legal representation and waiting for available court times. The dispute is decided by a neutral, third party with no necessary specialized expertise in the disputed subject. Legal procedures are highly structured with rigid, formal rules for presentations and fact finding. On the positive side, legal adjudication strives for consistency and predictability of results. The results of previous cases are published and can be used as precedents for resolution of new disputes.

6.8.4.4 Arbitration and Conciliation

Arbitration is a legal technique for the resolution of disputes outside the courts, wherein the parties to a dispute refer it to one or more persons (the "arbitrators," "arbiters," or "arbitral tribunal"), by whose decision (the "award") they agree to be bound. It is a settlement technique in which a third party reviews the case and imposes a decision that is legally binding for both sides.

In arbitration, the third party may make a decision that is binding on the participants. In conciliation, the third party serves only as a facilitator to help the participants reach a mutually acceptable resolution. Like negotiation, these procedures can be informal and unstructured.

6.8.5 Contract Termination

Early termination of a contract is a premature contract closure, resulting following a mutual agreement of both parties or from the default of one of the parties. It is imperative that the awarded contract describes clearly the rights and responsibilities of the parties in the event of an early termination.

6.8.5.1 Termination for Convenience

The buying organization may have the right to terminate the whole contract or a portion of the project, for cause or convenience, at any time (for example, if the project is cancelled). Clauses to this effect must be in place when the contract was awarded. Often the vendor will be compensated for any ongoing, completed, and accepted work related to the terminated part of the contract.

6.8.5.2 Default of Contract

Early termination can be called by the buying organization when the contractor fails to perform any provision of the contract including: failure to deliver by scheduled date; failure to maintain costs within an agreed threshold; failure to make progress (endangers performance of the contract), or other default deemed to cause damage to the buying organization, such as misdemeanor.

Penalties to be paid by the vendor may be invoked, according to agreed clauses in the awarded contract.

Clauses in the awarded contract can also include default by the buying organization. These would also provide compensation to the vendor to be paid for by the procurer.

6.8.6 Exercise: Status of Contracts on Your Project

Review your current project and report on the status of the contracted products and services.

FIGURE 6.20
Process step—closeout

6.9 PROCUREMENT PROCESS: CLOSEOUT

The closeout step (see Figure 6.20) terminates the procurement plan for the product or service. Verification is performed to ensure that the procured products/services have been received and are acceptable; the status of any outstanding invoices is established and the final payment is made. Additionally, vendor performance and lessons learned are documented. The contract is closed when the project procurement plan has been completely executed.

6.9.1 Detailed Closeout Step Process

- Verify product and services meet acceptance criteria
- Complete fiscal activities:
  - Approve final payment (notify appropriate accounts payable office)
  - Update project and activity records
  - Close purchase order (notify appropriate procurement office)
  - Update contract file

- Produce vendor performance evaluation and lessons learned
- Archive contract file (include completed project procurement plan with Project Plan) (see Figure 6.21)

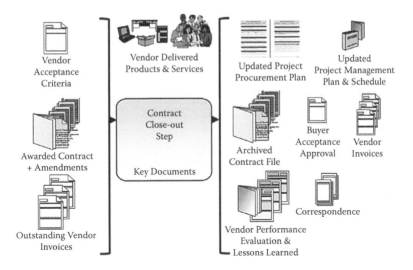

FIGURE 6.21
Key documents—closeout

6.9.2 Contract Closeout

Contract closure, which in some cases is a subset of the project closeout phase, is primarily focused on verifying that all products and services have been delivered according to the contracted acceptance criteria, including any changes incorporated during the vendor's implementation.

On satisfying the acceptance criteria, the project manager, or the contract administrator, provides the vendor with a formal written acceptance approval notice that the contract has been completed. The project manager and the vendor may agree that certain defects can be placed on a defect list or "punch list," to be resolved within an agreed schedule post-delivery. Unresolved claims may be subject to litigation after contract closure. In cases where the vendor holds multiple awarded contracts, each is closed independently.

Other activities performed by the project manager and the team members involve finalizing and updating vendor performance evaluation and other records that reflect final results have been achieved. Outstanding invoices are settled. These documents are archived along with the awarded contract files.

7

Project Implementation

7.1 CHAPTER OVERVIEW

This chapter covers in detail the project implementation phase from start of project execution to final acceptance, prior to progressing to the project close-out phase.

Special emphasis is placed on scope and change request management, accompanying the techniques for project performance tracking, control, and evaluation.

Team management is introduced, and will be further expanded in Chapter 9, Project Leadership Skills.

7.2 PROJECT IMPLEMENTATION OVERVIEW

The essence of the project implementation phase is to develop the product or service that the project was commissioned to deliver. Typically, this could be the longest or costliest phase of the project development life cycle, where most resources are applied.

This phase is focused on the execution and control of the approved project baseline, and utilizes and refers to all the project plans, schedules, and procedures, as well as the product requirements and design documents prepared during the initiation and planning phases.

The end of the phase is accomplished when the product or service is fully developed, tested, accepted, implemented, and transitioned to the performing organization.

Full and up-to-date documentation is to be kept throughout this phase, as this serves as input to the final phase, project closeout (see Figure 7.1).

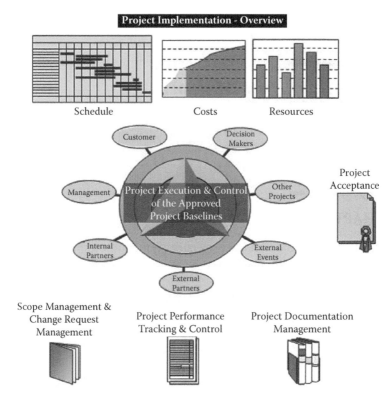

FIGURE 7.1
Project implementation overview

The overall management of the project implementation phase then revolves around the following themes:

- Scope management and change request management (discussed in Section 7.3.)
- Project performance tracking and control (discussed in Section 7.4.)
- Project documentation management (discussed in Section 7.6.)

The phase in completed with:

- Project acceptance (discussed in Section 7.7.)

The primary step of this phase is the project implementation start-up. This is an important platform for the project manager to formally begin the project implementation phase; institute the project team organization and management; and present, discuss, and review the current status and documentation of the project with team members.

7.2.1 Effective Project Implementation Start-Up

The purpose of the project implementation start-up (also called kick-off) is to formally acknowledge the beginning of the project implementation phase and facilitate the transition from project planning. The project implementation phase start-up ensures that the approved project is known to all participating team members and supporting entities, and is focused on the original business need. The start-up provides new team members a thorough orientation in preparation to their task responsibilities and performance. The project manager must be conversant in the current project documentation, which is reviewed with the meeting participants. Project goals, objectives, and all deliverables are presented, giving all team members a common understanding of the extent and scale of the project (see Figure 7.2).

7.2.1.1 Conducting the Implementation Start-Up Meeting

The project implementation start-up is a meeting conducted by the project manager, who presents the main components of the project plan. The goal

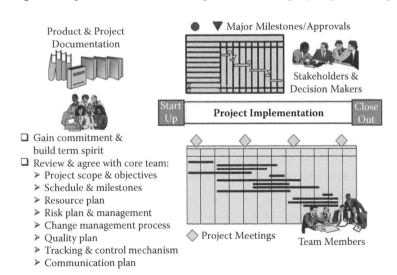

FIGURE 7.2
Implementation start up

of the meeting is to verify that all parties involved have consistent levels of understanding and acceptance of the scope of work to be performed, to validate expectations regarding the deliverables to be produced, and to clarify and gain understanding of the expectations and roles and responsibilities of each team member.

Core participants at the project implementation start-up meeting include the project team, the project sponsor, and key stakeholders. This is an opportunity for the project manager, assisted by the project sponsor, to emphasize the importance of the project and how it supports the business need.

The start-up meeting may be informal when the size of the project and the team are small, and the meeting can take place, for example, at the project manager's desk. For projects of greater size and/or importance, a formal meeting with a structured agenda and official invitations to participants is more appropriate. When the project utilizes subcontractors, the start-up may be performed in two stages, the first internal to the performing organization and the second with the contractors. For implementation start-up of commercial projects, formalism is highly recommended with key representatives of the client/customer organization.

The start-up meeting should be an event and needs careful preparation by the project manager. A suitable location should be chosen, preferably off-site, with provision for audio/visual facilities. The event duration will greatly depend on the nature and size of the project and the list of attendees. The key is to allow the appropriate time and not dash through the meeting, which may indicate that the project is not that important.

A suggested agenda:

- Introduction: Welcome address and presentation of attendees
- Project position in the overall business need
- Project overview:
  - Goals, objectives, key milestones, critical success factors
  - Stakeholders
  - Project organization: Roles and responsibilities of each team member
- Project description:
  - Scope and deliverables
  - Schedule and cost baselines
  - Milestones
  - Resource plan

- Project plans, processes, methods, and tools:
  - Scope and change management
  - Risk plan
  - Quality plan
  - Communication plan
  - Acceptance plan
- Tracking and control mechanism
- Meeting summary and action items
- Team-building activity
- Start-up meeting closure

7.2.1.2 Project Plan Considerations for Implementation Start-Up

At the project implementation start-up meeting, the project manager has an ideal platform to develop the philosophy to adopt during project execution and control. Pragmatism and realism have to be emphasized in conjunction with innovation and creativity in order to overcome a standard and stereotyped view that project scope, schedules, and estimated costs are static in a deterministic environment with complete information.

All project plans are not perfect because of their inherent characteristics: incomplete scope, assumptions, insufficient WBS detail, estimation error ranges, risks to manage, and most of all changes outside of the control of the project. The project manager must thus support proactive and reiterative project plan scheduling that anticipates and reacts to evolution and changes that are expected to occur during project execution.

The most common approach to project plan management is the rolling-wave planning technique (see Figure 7.3). This technique views the approved project plan as a strategic macro plan to reach the project goals, and breaks the total project elapsed time window into smaller tactical and detailed periods. During these shorter duration periods, tasks are assigned and work is performed. The choice of tactical plan periods is discretionary. It is common to make these on a four-weekly basis. For projects of a short total duration, weeks can be preferred. Long duration projects will usually be broken down into quarterly periods.

The tactical plan is the focus period for the monitoring, tracking, and control of work performance, risk and change request management, and progress reporting. Derived from the latest project plan, the tactical plan is established: further breakdown of tasks; specific work assignments are

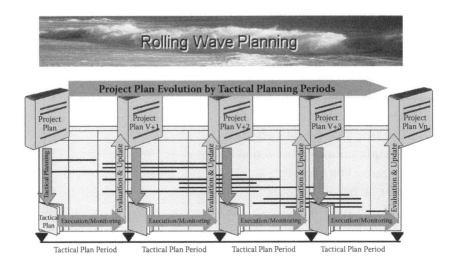

FIGURE 7.3
Rolling wave planning

made; reestimation of tasks is performed where necessary; contingency plans are detailed; and change requests are implemented. At the end of the tactical plan period, the project is evaluated and the strategic project plan is updated to reflect its latest and actual status. This process is repeated until the last tactical plan period is completed, thus completing the project.

7.2.1.3 Specific Focus Areas for Implementation Start-Up

The project manager must promote awareness, understanding, and the "big-picture" view to all engaged in the project implementation.

Obviously the business benefits derived from the project are to be emphasized. The project's major constraints need to be underlined, as well as the intermediate "must-dates" and interfaces to other projects, as they will be challenged throughout the project implementation phase, as real and actual results are confronted by the expectations, needs, and goals of the project. In the case the performing organization follows a project portfolio management system, the project's priority must be well understood as well as which other projects would be contending for limited funding and resources.

Space and time is to be allocated during start-up to reassess the scope of work and all commitments, both from inside the organization for such things as resource availability, and from external providers identified in the procurement management plan.

The project manager must seek to promote open discussion and the freedom to raise issues, as and when they occur. Project scope, schedule, and cost constraints are sufficiently coercive for the project manager not to instill an additional management-by-fear atmosphere, where no errors are admitted, variations are not acceptable, and problems are unwelcome. Too often team members performing in such an environment would adopt a "volcano syndrome" attitude, by choosing to sit on problems to stifle or cover them up hoping they will either disappear or never be noticed. Unfortunately, Mother Nature's volcanoes burn the backsides of problem sitters before they explode.

Organizational readiness is another area to highlight during the implementation start-up meeting. Team members and representatives of the performing organization will need to discuss and develop at this early stage the plans for the end of phase acceptance and hand-over. By using the rolling wave planning technique, major activities and milestones can be set to anticipate and prepare for the transition of the delivered product/ service to the operational environment.

7.2.2 Project Team Management

Chapter 9, Project Leadership Skills, offers an extensive discussion on project management leadership and team management. However, in brief, the primary objective of project team management is to improve overall project performance by enhancing team members' abilities to contribute quickly and positively to the project's desired outcome. When team members are encouraged to perform to high standards and are motivated about a project, they will usually seek ways to improve their individual skills to be more effective in performing their assigned activities. Understanding the importance of interaction among team members will often increase the willingness to identify and proactively address conflicts.

The project manager must not neglect the following:

- Team recognition and rewards: Intended to promote, encourage, and reinforce desired behavior or exceptional performance. Recognition programs must be documented clearly so team members understand what level of performance warrants an award.

- Team-building activities: To specifically improve the performance of the entire team, team-building activities provide opportunities for team members to improve their interpersonal and working relationships.

7.2.3 Exercise: Project Implementation Start-Up Agenda

Develop, or review, an agenda for your project implementation start-up meeting.

7.3 SCOPE MANAGEMENT AND CHANGE REQUEST MANAGEMENT

7.3.1 Definition of Change

Change will happen in all projects. "If you say *no* to change, change will say *no* to you" is a good phrase to keep in mind, as one of the major paradoxes of the management of projects is that projects exist to create change in an environment of change.

The word "change" should be qualified in the different contexts in which it is applied and depending upon what is being changed (see Figure 7.4). The following definitions will be used in this module:

- **Scope change:** The scope management process, in which a request is considered to change the agreed scope and objectives of the project to accommodate a need not originally defined to be part of the project
- **Change request management:** The management process for requesting reviewing, approving, applying, and controlling changes to the project's deliverables. Change request management is initiated following the first version of a completed and agreed deliverable

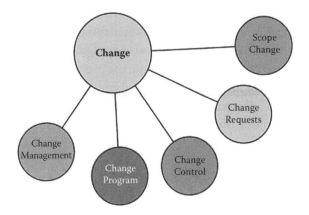

FIGURE 7.4
Types of changes

- **Change control:** Also called configuration management, is the technical and administrative control of the multiple versions or editions of a specific deliverable, after it was initially completed. Most typically this applies to objects, documentation, modules, and data definitions
- **Change management:** Normally refers to the organizational achievement of change in human behavior as part of an overall business solution
- **Change program:** The overarching organizational solution aligned to the strategic business intent to realize a business benefit (not confined to just the human behavioral element)

7.3.2 Scope Management and Change

Wisdom has shown that the prime focus for the project manager should not be to deliver the agreed scope on time and on budget. This may seem "blasphemy" for those project managers who religiously and blindly abide by the "Triple Constraint" philosophy pronounced by much project management literature. The most important objective is to optimize the benefit that change can generate for the project. The project manager should not have an attitude of resistance to all scope changes. This signifies that allowing the scope to change will enhance the project's benefit and subsequent operational use. When a scope change generates improved benefit, thus to be considered positive and not negative as most people might tend to concur, it should be proposed to the project's decision-making body.

However, and not contrary to what is written here, project managers should pay a great deal of attention to managing project scope and the changes made. Too many projects fail because of poor scope management. Allowing the project's scope to change during project implementation will generally augment costs, increase risks and lengthen duration. All may be justifiable. The pitfall is to allow a large number of "small" scope changes to be applied without impact analysis or control. These cause the most damage to the project and not the large and visible changes, which are by nature obvious to see by all.

Scope changes are a fact of project life and should be rigorously controlled to achieve satisfactory project delivery. The project manager should, however, be aware that many requested changes have ulterior motives. When external factors outside of the project's control are invoked, this may hide the fact that the initial approved scope was incomplete or incoherent. When providers and contractors press for certain scope changes, their aim may be to expand and extend their profit.

7.3.3 Scope Management at Project Implementation Start

An efficient scope management and change request management process has to be defined and agreed prior to project implementation. Project scope should have been clearly defined at project baseline approval, as potential changes are assessed and the project's performance is measured against the established initial baseline.

The process has to be flexible and adaptable, without losing its essence. If it is too burdensome, either valuable changes will not be requested or individuals will just circumvent it, provoking uncontrolled modifications to the project scope. If the process is too slack, then changes may be integrated without adequate consideration as to their value and consequence.

7.3.4 Change Control versus Issue Management

There are many similarities between change control and issue management. A large percentage of "issues" will directly or indirectly provoke change, whilst most changes echo issues.

This chapter suggests presenting issue management as a separate but related process whereby an issue can evolve into a change request where appropriate (see Section 7.4, Project Performance Tracking and Control).

7.3.5 Origin of Changes

For the project manager and the core team members, the principal approach to scope management and change request management is to promote an attitude that realistically and pragmatically addresses the areas and different origins of change and reacts in consequence (see Figure 7.5). The major origins of change to projects may be as follows:

- **Business needs:** Business needs change due to an effervescent and volatile market of keen competition, regulations, and environmental and social issues. All businesses must be willing to change if they are to remain competitive, and the key project stakeholders are in the front line of these changes.
- **Organizational restructuring:** Organizations refine and/or redefine their business models to maintain, sustain, and grow their operation. These changes, which cover both local and global operations, may involve functional departments, product lines, new business partners and channels, internal processes, new financial structures, expansion/reduction of locations, etc. Restructuring may also be invoked by mergers, acquisitions, or takeovers. The more the project is mapped to the company's strategic intent, the more will changes affect the overall performance and achievement of initial goals.

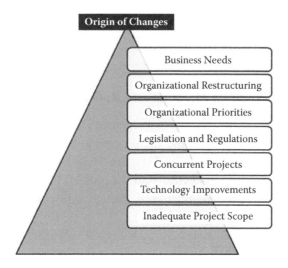

FIGURE 7.5
Origin of changes

- **Organizational priorities:** The market dynamics and the organization's responses may shift the importance and ranking of projects, while the project scope remains valid.
- **Legislation and regulations:** These external requirements will bring about forced changes to those projects exposed to decisions made by national and international regulatory bodies.
- **Concurrent projects:** Other interrelated projects and initiatives within the organization may impose changes to the project.
- **Technology improvements/innovation:** Changes resulting from internal innovations or from pressure from the market and the competition, which will force a redefinition of the current product specification and project.
- **Inadequate project scope:** Elements of the project's design and deliverables do not fully meet the defined need and changes will have to be made to realign the scope to the desired goals.

7.3.6 Exercise: Draft the Project Requested Changes and Their Origins

Assess your current project(s) and draft the changes that have been requested and their origins.

7.3.7 Change Request Management Roles

A key aspect in change request management is to allow individuals to raise changes, while channelling the process to be systematic with roles and responsibilities clearly defined. Figure 7.6 is a template for foremost resources (both within and external to the project) involved with the initiation, review, and implementation of changes within a project.

7.3.7.1 Project Manager

The project manager is closely concerned with the overall process and ensures that all change requests are handled effectively. In the majority of

Change Request Management Roles

Change Requestor

Change Manager

Project Manager

Change Analysis Group

Change Approval Group

Change Implementation Group

FIGURE 7.6
Change request management roles

cases, the project manager also acts as the change manager, as described in Section 7.3.7.3.

Special attention is given to stakeholder and client expectation management, scope of work, and contractual changes resulting from the request.

As no process can be perfect, the project manager should also monitor informal changes made outside of the process, and subsequently raise formal change request forms in order to reconstitute the flow of the process. Informal changes are usually discovered during project review meetings.

7.3.7.2 Change Requestor

The change requestor recognizes a need for change to the project and formally communicates this requirement to the change manager. The change requestor may be any individual associated with or beneficiary of the project, and is responsible for:

- The early identification of a need to make a change to the project
- The documentation of that need, through the completion of a change request form
- The submission of the change request form to the change manager for review

7.3.7.3 Change Manager

The change manager, usually the project manager, receives, logs, monitors, and controls the progress of all changes within a project. The change manager is responsible for:

- Having a close working relationship with the project manager (unless, of course, it is the same person).
- Receiving all change requests and logging those requests in the change register.
- Categorizing and prioritizing all change requests.
- Reviewing all change requests in order to determine if additional information is required prior to submission to the change review group.
- Determining whether or not a formal change impact analysis (feasibility study) is required in order to complete a change request submission.
- Initiating the change impact analysis, through assignment of the change analysis group.
- Developing the implementation schedule and cost for changes (as agreed by the change approval group). This is done in close collaboration with the project manager.
- Monitoring the progress of all change requests in order to ensure process timeliness.
- Escalating all change request issues and risks to the change approval group.
- Reporting and communicating all decisions made by the change approval group.

7.3.7.4 Change Analysis Group

The Change Analysis Group, usually members of the project team, completes formal analysis and feasibility studies for change requests issued by the change manager. The change analysis group is responsible for:

- Performing a research and analysis exercise in order to determine the likely options for change, costs, benefits, and impacts of change
- Documenting all findings within a change impact analysis report
- Performing a quality review of the report
- Forwarding the report to the change manager for submission to the change approval group
- Establishing the total costs incurred for the change impact analysis study

The project manager has to address the schedule conflicts that arise when the members of the change impact analysis group are the same as the project team. At issue is the schedule disruption caused by resources assigned on change impact analysis and not on ongoing planned tasks.

7.3.7.5 Change Approval Group

The change approval group, often called the change control, determines the authorization of all change requests forwarded by the change manager. The change approval group is usually constituted of the project sponsor and selected key stakeholders. For commercial projects, the change approval group is extended to the client/customer organization. The change approval group is formally responsible for:

- Having a close working relationship with the project manager
- Reviewing all change requests forwarded by the change manager
- Considering all supporting change documentation
- Approving/rejecting each change request based on its relevant merits
- Resolving change conflict (where several changes overlap)
- Resolving change issues
- Determining the change implementation timetable (for approved changes)

7.3.7.6 Change Implementation Group

The change implementation group, usually members of the project team, performs and reviews the implementation of all changes within a project. The change implementation group is responsible for:

- Having a close working relationship with the project manager
- Organizing detailed scheduling of all changes (subject to the general timeframes provided by the change approval group and the project manager overall schedule)
- Testing of all changes, prior to implementation
- Implementing all changes within the project
- Updating all documentation affected by the change
- Reviewing the success of a change, following implementation
- Requesting closure of a change within the change log

The project manager has to address the schedule conflicts that arise when the members of the change implementation group are the same as the project team. At issue is the schedule disruption caused by resources assigned on change implementation and not on on-going planned tasks.

7.3.8 Exercise: Change Request Roles and Responsibilities

Assess your current project(s) and list the individuals or groups who have any role and responsibility in the change request process of your company.

7.3.9 The Change Request Management Process

The change request management process combines the procedures and responsibilities to effectively execute a well-controlled and efficient process. The process spans the steps starting from the original request through to the application of the approved change to the final decision. The change request management process binds scope change management with change control where all key players fulfill their roles (as explained previously).

The change request management process (see Figure 7.7) revolves around five key processes:

- **Step 1—Record change request:** Submission, receipt, review, and logging of change requests
- **Step 2—Change impact evaluation:** Conducting the change impact analysis and producing an impact evaluation
- **Step 3—Change impact proposal:** Submitting the change impact evaluation for approval
- **Step 4—Change request approval:** Reaching decisions for change request approval
- **Step 5—Change request implementation:** Applying decision and closure of change request

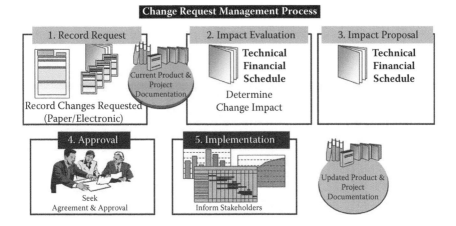

FIGURE 7.7
Change request management process

Documentation of the "life" of a change request is maintained throughout all the steps, as this will provide a trace and allow substantiation of the project's evolution in scope, duration, and cost.

7.3.9.1 Step 1—Record Change Request

This step initiates the whole process, providing the ability for the change requestor, any person associated to or benefitting from the project, to submit a request for change to the project. The change request form (CRF) is the mechanism for recording and controlling all project changes (see Figure 7.8).

Part 1—Initiating is used for change request submission and is logged in the change register with a unique change ID.

Part 2—Impact analysis summary is completed following the change impact analysis, and is the cover sheet to the detailed documentation submitted to the change approval group (see next step).

Change request forms are recorded in the change request register (see Figure 7.9). The register is maintained by the project manager, providing easy summary reference to the number of change requests and their status.

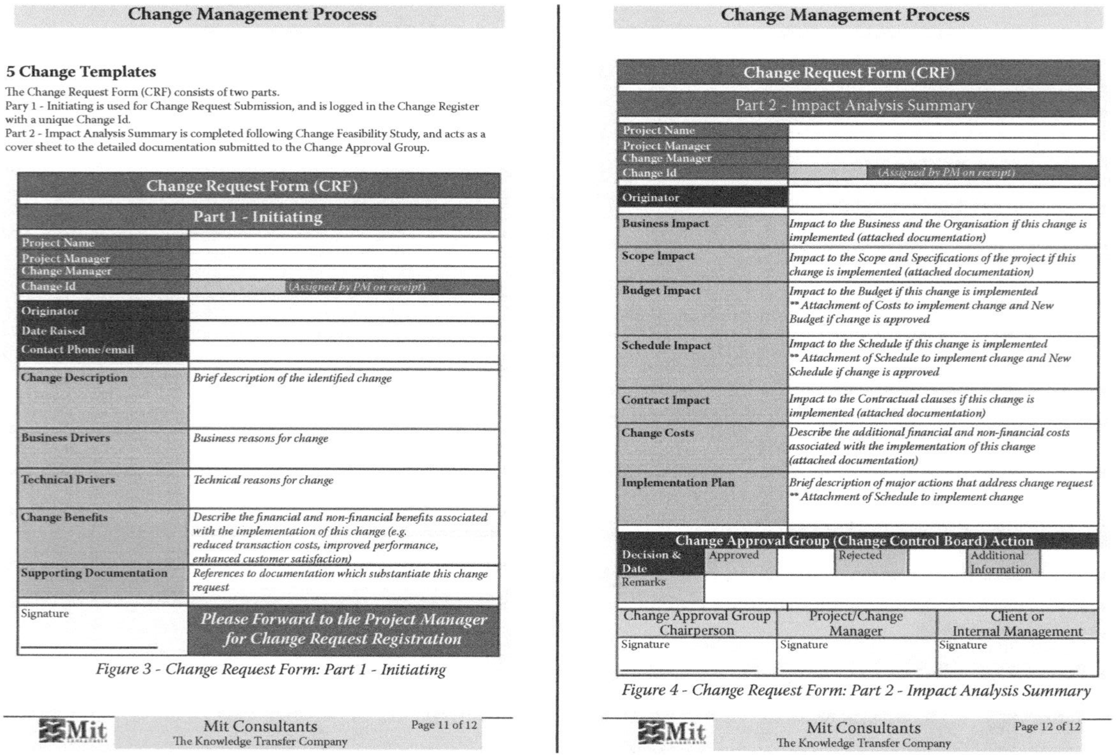

Change Management Process

5 Change Templates

The Change Request Form (CRF) consists of two parts.
Pary 1 - Initiating is used for Change Request Submission, and is logged in the Change Register with a unique Change Id.
Part 2 - Impact Analysis Summary is completed following Change Feasibility Study, and acts as a cover sheet to the detailed documentation submitted to the Change Approval Group.

Change Request Form (CRF)	
Part 1 - Initiating	
Project Name	
Project Manager	
Change Manager	
Change Id	*(Assigned by PM on receipt)*
Originator	
Date Raised	
Contact Phone/email	
Change Description	*Brief description of the identified change*
Business Drivers	*Business reasons for change*
Technical Drivers	*Technical reasons for change*
Change Benefits	*Describe the financial and non-financial benefits associated with the implementation of this change (e.g. reduced transaction costs, improved performance, enhanced customer satisfaction)*
Supporting Documentation	*References to documentation which substantiate this change request*
Signature	***Please Forward to the Project Manager for Change Request Registration***

Figure 3 - Change Request Form: Part 1 - Initiating

Mit Consultants
The Knowledge Transfer Company
Page 11 of 12

Change Management Process

Change Request Form (CRF)	
Part 2 - Impact Analysis Summary	
Project Name	
Project Manager	
Change Manager	
Change Id	*(Assigned by PM on receipt)*
Originator	
Business Impact	*Impact to the Business and the Organisation if this change is implemented (attached documentation)*
Scope Impact	*Impact to the Scope and Specifications of the project if this change is implemented (attached documentation)*
Budget Impact	*Impact to the Budget if this change is implemented ** Attachment of Costs to implement change and New Budget if change is approved*
Schedule Impact	*Impact to the Schedule if this change is implemented ** Attachment of Schedule to implement change and New Schedule if change is approved*
Contract Impact	*Impact to the Contractual clauses if this change is implemented (attached documentation)*
Change Costs	*Describe the additional financial and non-financial costs associated with the implementation of this change (attached documentation)*
Implementation Plan	*Brief description of major actions that address change request ** Attachment of Schedule to implement change*

Change Approval Group (Change Control Board) Action						
Decision & Date	Approved		Rejected		Additional Information	
Remarks						

Change Approval Group Chairperson	Project/Change Manager	Client or Internal Management
Signature	Signature	Signature

Figure 4 - Change Request Form: Part 2 - Impact Analysis Summary

Mit Consultants
The Knowledge Transfer Company
Page 12 of 12

FIGURE 7.8
Change request form

Change Request Register										
Project Name:										
Project Manager:										
Change Manager:										
Summary			Description			Approval			Implementation	
Change ID	Date Raised	Raised By	Description of Change (can refer to a document)	Impact (L/M/H/X)	Approved By	Approval Date	Status (Go/NoGo)	Implementation (can refer to a document)	Completion Date	Review Date

FIGURE 7.9
Change request register

Part 1 of the CRF is raised and submitted (see Figure 7.10):

- Change requestor identifies a requirement for change to any aspect of the project (e.g., scope, deliverables, timescales, and organization).
- Change requestor completes a CRF and forwards the form to the change manager. The CRF provides a summary of the change required, including the:
  - Change description;
  - Reasons for change (including business drivers);
  - Benefits of change;
- Supporting documentation.

The CRF consists of two parts.

Not all changes, however, follow the overall approved process and many do not enter it by this initial step. This often occurs when project team members may be persuaded to apply changes, however minor they may seem, without using the approved procedure. Although this can seem practical, the project manager should be alert for uncontrolled changes. Where necessary, applied changes should be channelled into the process and be recorded.

Change Request Form (CRF)	
Part 1 - Initiating	
Project Name	
Project Manager	
Change Manager	
Change Id	*(assigned by PM on receipt)*
Originator	
Date Raised	
Contact Phone/e-mail	
Change Description	*Brief description of the identified change*
Business Drivers	*Business reasons for change*
Technical Drivers	*Technical reasons for change*
Change Benefits	*Describe the financial and non-financial benefits associated with the implementation of this change (e.g., reduced transaction costs, improved performance, enhanced customer satisfaction)*
Supporting Documentation	*References to documentation which substantiate this change request*
Signature _____	***Please Forward to the Project Manager for Change Request Registration***

FIGURE 7.10
Change request form, part 1

Before proceeding to the next step, for change impact evaluation, a change request type or severity should be assigned:

- New/Additional function: A new or additional specification not identified in the initial project scope
- Enhancement: An alteration to an existing function that has been identified in the initial project scope
- Legal or regulatory: Mandated requirement not known or missed during the initial project scope
- Critical defect: Blocks project implementation
- Deficiency in project scope:
  - Major—Functional or design defect
  - Minor—Reduced functionality or performance
- Trivial—Cosmetic problem

7.3.9.2 Step 2—Change Impact Evaluation

This step consists of evaluating and preparing a full change impact report, in order to ensure that all change options have been investigated and presented accordingly. The change impact analysis will consist in defining the:

- Change requirements
- Change options
- Change costs and benefits
- Change risks and issues
- Change impact
- Change recommendations and plan

The change impact analysis can be a time-consuming and costly exercise. The project manager should assess how resources will be assigned to this task without jeopardizing the on-going implementation. The detailed impact analysis should address how, what, and where the proposed change will affect:

- Conformance to the current project scope with respect to physical, functional, quality, and agreed design
- Use of existing infrastructure, such as equipment, machinery, computer systems, etc., currently operated or planned for the project
- Products being developed in parallel, or in associated projects
- The impact to the business
- Later stages in the project and/or operations post-delivery
- Post-implementation support levels to the users and/or clients
- Deliverables already complete
- The technical skills, training, or specialist advice required to resolve the problem, or to implement the change
- Other areas, if any, in quality procedures, testing, operating procedures, and user training

A review of the change impact analysis is then performed before proceeding to the next step—change impact proposal.

7.3.9.3 Step 3—Change Impact Proposal

This step consists of collating the documentation raised from the original change request to that from the impact analysis. The whole set is then submitted to the change approval group for review and approval. This documentation includes:

- The original CRF
- The approved change impact analysis report
- Any supporting documentation

Part 2 of the CRF is the cover sheet of the documentation package (see Figure 7.11).

Change Request Form (CRF)		
Part 2 - Impact Analysis Summary		
Project Name		
Project Manager		
Change Manager		
Change Id		*(assigned by PM on receipt)*
Originator		
Business Impact	*Impact to the Business and the Organisation if this change is implemented (attached documentation)*	
Scope Impact	*Impact to the Scope and Specifications of the project if this change is implemented (attached documentation)*	
Budget Impact	*Impact to the Budget if this change is implemented ** Attachment of Costs to implement change and New Budget if change is approved*	
Schedule Impact	*Impact to the Schedule if this change is implemented ** Attachment of Schedule to implement change and New Schedule if change is approved*	
Contract Impact	*Impact to the Contractual clauses if this change is implemented (attached documentation)*	
Change Costs	*Describe the additional financial and non-financial costs associated with the implementation of this change (attached documentation)*	
Implementation Plan	*Brief description of major actions that address change request ** Attachment of Schedule to implement change*	
Change Approval Group (Change Control Board) Action		
Decision & Date	Approved Rejected	Additional Information
Remarks		
Change Approval Group Chairperson	Project/Change Manager	Client or Internal Management
Signature	Signature	Signature

FIGURE 7.11

Change request form, part 2

In certain cases, especially when the requested change required extensive feasibility analysis, the project manager will also record the costs associated to the change impact analysis, as these costs may well increase significantly the total budget. For commercial/external client projects, and according to pre-established terms and conditions clause, the change impact analysis may constitute additional invoicing.

7.3.9.4 Step 4—Change Request Approval

This step consists of presenting the change impact analysis report for formal review by the change approval group and seeking a decision to pursue, reject, or suspend the change request. The change approval group will choose one of the following outcomes regarding the change proposed:

- Approve the change as requested
- Approve the change subject to specified conditions
- Reject the change
- Request more information related to the change
- Suspend the request

Further to the actual contents of the change proposal and its business merits, the change approval group should consider:

- Risk to the project in implementing the change
- Risk to the project in *not* implementing the change
- Impact to the operation in implementing the change (additional time, resources, funding)

Special attention should be given to changes affecting an external provider, as these would need to be reviewed with that contractor to agree on any necessary contract revisions or payments, etc. Changes of scope and contract revisions would subsequently require the approval of the steering committee (or it might have been a change control board).

The change request management process should define the constitution and the decision-making approval responsibilities, and authority levels of the change approval group, so that routine changes can be approved and implemented rapidly by members of the performing team, whilst those of a greater importance are reviewed and approved by higher level management, and when necessary by the client/customer. The latter approval cycle would require more time before the requested change can be implemented.

For example, when the project scope is impacted by a proposed change that redefines the business outcome and benefits of the project, the change approval group is to be formed by the sponsor and key stakeholders (business owners, steering committee, client/customer, etc.) as the change requires a business decision. When the major vectors of the project are not affected, such as no changes to the baseline and contracted delivered scope or no variations outside the accepted and agreed ranges for schedule and cost, the change approval group can be constituted by the project manager and selected team members, who are given authority to approve the change within certain limits.

A more formal way for approving changes is to institute a change control board (CCB) as the change approval group. This will usually consist of the sponsor and key stakeholders, and may be extended to include subject-matter experts. For large complex projects, there may be different change control boards reviewing different areas of the project: for example middle and upper managers for business process changes, technical experts for technology-related changes, and functional and HR managers for organizational issues. The project manager is to schedule CCB review meetings on a regular basis to review and approve changes, as in most cases, the CCB cannot be available as and when required nor for long periods of time.

Finally, the change approval group may be a steering committee. The project manager must establish with the sponsor the nature and level of impact of the change, above which the change request approval would be referred to the steering committee.

7.3.9.5 Step 5—Change Request Implementation

This step concerns the complete implementation of the change.
This includes:

- Identifying the change schedule (i.e., date for implementation of the change)
- Testing the change prior to implementation
- Implementing the change
- Updating all project documentation
- Reviewing the success of the change implementation

- Communicating the success of the change implementation
- Closing the change in the change log

The project manager will also ensure that any contractual clauses in the terms and conditions are updated accordingly. Additional invoices, when applicable, are raised for the implemented changes.

After change approval is given, the work is assigned to the appropriate project team members and/or the providers and subcontractors. When change implementation is complete, the results are reviewed and the change request closed. It is at times preferable to implement the change request in more than one stage. For example, it may be better to introduce a temporary solution so that the overall benefit from the project can be delivered, and then build a permanent solution after the system is live.

7.3.10 Contractual Consequences

Particular considerations occur where changes impact work performance with an external provider or subcontractor. Changes emanating from the performing organization and requested to the provider would require negotiations on the original contractual terms and adjustments to both the schedule and cost/price agreements. The performing organization would then be responsible for any additional costs and project delivery impediment.

Conversely, when changes are brought upon by the provider, the impacts to the project delivery have to be measured to ascertain if and when schedule slippages or supplementary costs will invoke contractual clauses, such as penalties or other damages. Negotiations will need to refer to the original type of contract established at contract award.

7.3.11 Project Plan Update

The project plan is updated according to the frequency established either by the project review/progress schedule or by the rolling wave planning milestones.

When variances are within the limits set by the project sponsor, the project manager and the core team members proceed to make the necessary adjustments and publish the new project plan. If, however, the limits are exceeded, then a proposed adjusted project plan is presented to the project sponsor for review and approval.

Pending change requests may be deferred for processing to the next tactical plan period, or at a later period, depending on the current state of the project and the importance of the requests. This would be an option for changes that are requested in the closing stages of delivery and that may disrupt the completion of the project. It might also be nonbeneficial to delay the entire project to accommodate a change that may cause unnecessary delays that jeopardize the core project functionalities to be introduced. Similarly, certain change requests may be suspended during the project implementation to be effected during the support stage after the project is complete.

At the end of the project, pending change requests and outstanding actions, resulting from "punch lists," are reviewed and the appropriate procedure is initiated to get them addressed when the delivered product/service is in operational use.

7.3.12 Configuration Management

Configuration management is the overarching process to manage the status and evolution of the project and its documentation. It embraces scope and change request management, documentation, and quality management.

Project configuration management (PCM) is the collective body of processes, activities, tools and methods project management practitioners can use to manage items during the project life cycle. PCM addresses the composition of a project, the documentation defining it and other data supporting it. It is a baseline- and requirements-management process that provides managed control to all phases of a project life cycle.

7.3.13 Exercise: Your Change Request Management Process

Describe the change request management process used in your organization, and identify enhancements and improvements you can propose to render it more efficient.

7.4 PROJECT PERFORMANCE TRACKING AND CONTROL

7.4.1 Baseline Management

Baseline management identifies, establishes, documents, and controls baselines relevant to a project. It is used to measure how project performance deviates from the approved project plan.

The project's baselines are an integrated set of technical/performance, schedule and cost requirements. These are approved and documented before the project implementation execution and control activities can begin. The project baselines serve as the "contract" between the project sponsor and the project manager and team responsible for realizing the product or service.

At start of execution, the project's baselines are placed under change control in order to coordinate, approve, and document changes to the performance, schedule, cost, and benefit baselines during solution implementation. As seen previously, approved and implemented change requests redefine the baseline as the original plan plus the approved change.

7.4.1.1 Manage Project Schedule

In managing the project schedule, the project manager has responsibility to complete the project according to the established and approved timeline by ensuring that project tasks are performed accordingly to meet milestones and produce the project deliverables. The project schedule reflects any extensions as per approved and implemented change requests.

The actual performance and progress of project activities and tasks constitute the key tracking and control measures that are evaluated and compared to the current project schedule baseline.

7.4.1.2 Manage Project Budget

In managing the project budget, the project manager has responsibility to complete the project within the allocated and approved budget, plus extensions as per approved and implemented change requests.

Project budget management is concerned with all costs associated with human resources, equipment, materials, travel, and supplies. Increased costs of any type of resource, have a direct impact on the budget. The actual costs are tracked against estimates. Changes to the scope of the project will most often directly impact the budget.

The project manager's decision-making authority on budget decisions must be established and agreed with the project sponsor at baseline approval. For example, agreement is made to define whether the project manager has direct authority to allocate additional resources or must seek approval prior to any additional allocations.

The project manager will focus principally on the following project budget related elements: (see Section 7.5, Project Earned Value Management for details)

- The original estimated budget (cost) that was approved by the project sponsor
- The total cost of approved changes as a result of change control
- The most current approved project budget (the original estimated budget plus and the total cost of approved changes)
- The actual cost accrued to date on all tasks and resources in the project
- The estimated cost to complete the project tasks
- The forecast revised total cost of the project

7.4.2 Project Tracking and Control

The project team members use the communications mechanisms documented in the communications plan to provide feedback to the project manager on their progress at the activity or task level. The most common instrument is a project progress report on which is recorded the start and end of work performed, or the progress percentage completed for initiated and as yet unfinished tasks (see Figure 7.12).

Each organization pursues different ways to capture the actual costs incurred for the project. These vary from a task by task cost allocation to a blanket monthly cost report for the project provided by the financial department. The former is more meaningful for cost control and forecasting, whereas the latter is at times meaningless for activity/task cost control.

Data collected on actual performance, with additional information from the change request management process, is posted by the project manager against the current project schedule and budgeted cost curves. This should

Project Tracking & Control

Team Member
Progress Report

Deviations
to Schedule

Deviations
to Budget/Cost

Deviations
to Milestones

Change Requests

Risk Register

Project Manager
& Team Leaders
Meetings

Schedule Status

Budget/Real Costs

FIGURE 7.12
Project tracking and control

be done by the project manager on a regular basis and at least once a week to be able to respond quickly to variations.

7.4.3 Project Monitoring and Evaluation

Project monitoring is best done at the task level at a frequency compatible with the nature and duration of activities/tasks. There is no value in monitoring task performance on a daily basis for activities that have an average duration estimated in weeks, not to monitor performance on a weekly basis for activities/tasks of an average duration estimated in days.

Project evaluation is usually performed on a regular calendar basis, again compatible to the average estimated durations of activities/tasks. The most common frequency is a weekly evaluation. However, and to be aligned to management reporting, evaluation is often performed monthly. Evaluation consists of comparing actual schedule and cost progress against planned, and taking the appropriate measures to analyze variances and correct the project deviations where and when possible.

From the updated project schedule and cost curves, the project manager reviews the status of the project with the core team members. Areas to focus on include:

- Start/end of activities and percentage complete
- Task duration deviations
- Total project schedule deviations
- Status of deliverables completion
- Cumulative actual costs deviations to budgeted costs
- Total project cost deviations
- End of project schedule variation impacts
- Total project budget variations and funding requirements
- Status of risks (see Section 7.4.4., Risk Control)
- Status of raised issues (see Section 7.4.5., Issue Management)
- Resource allocation and forecast
- Project team members work load allocation and morale

When updating the project schedule, the project manager maintains the current schedule, and previous versions of the schedule are archived—thus creating an audit trail and history of the project.

Updating the project schedule is not limited to recording resource hours (or other time units). The percentage complete of the activity/task is also posted to bring coherence in appreciating the schedule variations. The schedule update will frequently require activities/tasks to be re-planned. When the updated project schedule results in a project delivery outside a previously agreed range, a new baseline schedule must be created and the project manager should obtain sponsor approval.

Positive or negative variances to the schedule and cost baselines are analyzed to take the appropriate proactive measures. The project manager needs to understand the variance causes and take the necessary corrective actions to regain control of the project schedule and budgeted cost. For example, additional funding or resources may be needed than originally planned for.

It is the responsibility of the project manager to ensure the accuracy of the updated project schedule as the principal method for managing the budget. This is of utmost importance to communicate exact project schedule and budget status, impact of changes, estimates to complete, and variance. This project manager will evaluate the project at different levels: activity/task, milestone, phase, resource, and deliverable.

7.4.4 Risk Control

The purpose of the steps to monitor and control risks is to deploy the risk management plans prepared in prior phases, and to develop and apply new response and resolution strategies to unexpected events.

Risks are potential future events that affect a project's scope, schedule, and/or cost. The risk response plans describe the actions that have been planned to mitigate and react to these potential events. As the project approaches the event dates, the project manager and the team members will conduct a risk re-evaluation of probability and impact for events in the risk register, as well as identify additional risk factors and events.

7.4.4.1 Monitor Risks

The project manager monitors the status of risk events recorded in the risk register, reassessing current events, inserting new risk events as they appear, reevaluating risk mitigation plans, and reviewing contingency plans. The risk register is reviewed and priority/ranking is re-determined, as the risk probabilities may have evolved; the expected impacts may be dissimilar. Consequently, the risk management plan is reevaluated.

7.4.4.2 Control Risks

Risk events on the risk register will occur. The project manager and project team members evaluate the risk occurrence and invoke the corresponding risk response:

- The risk event occurred as expected, and the existing risk response may be adequate.
- The risk event occurred not as expected, and the existing risk response may require modification.
- The risk event was unexpected and unanticipated, and a risk response must be developed to address it.

Irrespective of the scenario, however, when the risk event occurs it ceases to be a risk (i.e., a future and potential event) and is instead managed as an issue—a current and definite condition. See Section 7.4.5., Issue Management.

7.4.4.3 Monitor Risk Impact on the Baseline

During the risk control process, the project manager evaluates the effects on the project's scope, schedule, and cost. Frequently, risk events will occur without affecting (either positively or negatively) the project's baselines. However, when a risk event occurs that threatens any of the project baselines, the project manager must determine the appropriate action to protect the project.

7.4.5 Issue Management

Managing issues involves documenting, reporting, escalating, tracking, and resolving problems that occur as a project progresses. During project planning, the project manager and project sponsor agreed upon and documented the process for managing issues and included the process in the project plan. This process describes how issues will be captured and tracked, how issues will be prioritized, and how and when issues will be escalated for resolution.

Issues are usually problems and questions raised by members of the project team, including customer and contractors. Issues also include risk events that have occurred.

Issues are different, although related, to change requests, in that if they are within the limits of a predefined tolerance, they do not usually have an immediate impact on the project scope or schedule. Issue management is a reactive approach, which aims at resolving issues as and when they occur. Unresolved issues, however, will affect the project schedule or budget, triggering the need for change control.

Issues are recorded, and action plans are elaborated describing the necessary tasks to determine the impact to the baselines. Once the impact is ascertained and quantifiable, the project manager prioritizes the issue in relation to all other open issues. Highest-priority issues are addressed first, and the project manager invokes the change control process to resolve the issue.

The issues log includes the date the issue is recorded, the actions to be taken, the estimated resolution date, and the individual assigned to perform the actions (see Figure 7.13).

Project Issue Log			
	Project Id.	Project Description	Date

Issue No.	Reported By	Date of Entry	Issue Description (referencing corresponding project documentation)	Priority (H, M, L)	Status	Date Resolved	Resolution & Comments

FIGURE 7.13
Project issue log

Open issues are reviewed, as an agenda item, by the project and the team members in project status meetings. Unresolved issues can slowly cause project failure. The project manager must persistently pursue issue resolution. The project manager posts progress on the resolution of an issue, and a description of how the issue was resolved is recorded when it is closed.

7.4.6 Manage Acceptance of Deliverables

The acceptance management process, developed during the project planning phase, describes how deliverables will be reviewed and accepted according to the criteria developed for each. The acceptance management process is part of the project plan and documents:

- The description and definition of the deliverable submitted for acceptance.
- The criteria that must be met for each deliverable to be approved acceptable.
- The individuals from both the performing organization and the client/customer organization nominated to review each deliverable. In some cases, external independent subject matter experts may be involved for verification and validation of the deliverables.

- The individuals from both the performing organization and the client/customer organization who have the authority to sign the approval form, indicating acceptance.
- The time-frame during which deliverables must be either approved or rejected by the reviewers and approvers, and the cycle process for resubmission.

The acceptance management process is performed throughout the project implementation phase. Acceptance is intimately associated with client/customer expectations, and the project manager should constantly strive to understand and align the deliverables to these requirements.

Acceptance begins when the project manager, after formally reviewing the deliverable with the performer, presents a completed deliverable for acceptance. In many cases and for projects deemed to be "small" and/or not consequential to the overall business benefits, the same person may review, approve, and sign-off the deliverable acceptance.

When formalism is required and when contractual reasons dictate it, the deliverable is submitted first to the reviewer. After analysis, in accordance with the established acceptance criteria, the reviewer prepares a recommendation as to whether to accept or reject the deliverable. For the latter, the reviewer provides the reason for the refusal and transmits this to the project manager for action and/or correction. The review cycle must respect the preestablished time-frame indicated in the acceptance management process. There may be cases when many individuals participate in the review process in a planned sequence.

From the acceptance recommendations provided by the reviewer, the approver has the final decision to accept or reject the deliverable. This decision process must also be aligned with the agreed time-frame. If the approver decides that the deliverable be rejected, this is forwarded to the project manager with the reasons for rejection. Corrections and adjustments are appropriately made and the project manager resubmits the deliverable into the reviewer/approver cycle.

Formal review and approval of any key deliverable is a major elapsed-time issue that the project manager must manage tightly. Significant delays in the process should trigger the project manager to escalate the situation. Similarly, the project manager must track the number of times a deliverable is resubmitted. Following multiple rejections of the same deliverable, the project manager should take action to analyze the situation, resolve the issue, and, when necessary, exercise the appropriate escalation procedure.

Once a deliverable is considered acceptable, the project manager should obtain the appropriate signatures, which indicate formal acceptance of the deliverable.

7.4.7 Manage Project Transition

The project transition management plan, part of the project plan, is focused and dedicated to transitioning and/or deploying the product/service to the performing organization. This is sometimes called hand-over, ramp-up, roll-out, parallel running, etc. It focuses on the readiness of the performing organization to be prepared, trained and operationally able to exploit, use, utilize and sustain/maintain the delivered product/service.

Project transition includes:

- Coordinating and performing client/customer acceptance testing.
- Ensuring timely completion of all facilities issues, such as securing physical space, installing appropriate equipment, obtaining the appropriate building permits, etc.
- Managing organizational readiness activities that ensure the performing organization's capability to operate under the new situation to be delivered. This covers principally customer training and orientation, and can extend to new organizational and job definitions. The project manager must coordinate these efforts with the performing organization's change management plan.
- Managing the actual transition, according to a preestablished philosophy: from a phased, stepped transition, as for a progressive office move, to a one step "switch-over," as for an overnight change from an old to a new processing system.
- Ensuring supportability of the product/service in readiness for its operational use. This covers from the institution of internal "help" lines to specific support from external organizations with appropriate service level agreements.
- Produce and distribute all necessary documentation that will be provided with the product/service delivered. This will include:
  - User manuals
  - Online help
  - Assembly or usage instructions

The project manager must ensure that the transition plan and schedule is fully integrated into the implementation plan and schedule and coordinate this with the receiving performing organization.

7.4.8 Project Reporting

During project implementation, the project manager, project team, sponsor and stakeholders will share information using a variety of communication mechanisms (see Figure 7.14). These have been defined in the project communications plan during project planning and will principally include:

- Status meetings
- Meeting notes
- Status and progress reports
- Executive meetings
- Steering committee meetings

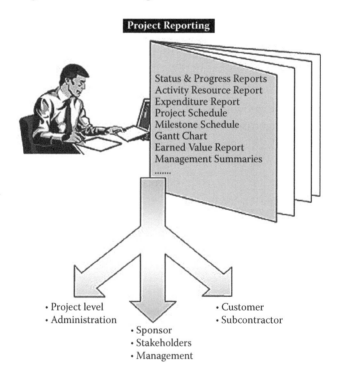

FIGURE 7.14
Project reporting

The project manager has the overall responsibility for project information collection, storage, and dissemination, and how the communications plan is articulated. As the project implementation progresses, the communications plan may require to be adapted and changed because of new events that vary the way information is accessed and disseminated. Even though the project manager is central in the collection and distribution of project information, communication is to be bidirectional with the project team and corresponding stakeholders.

Furthermore, the project manager should ensure that progress reports, meeting minutes, etc., are presented and written in a clear, unambiguous, and understandable manner.

7.4.9 Status, Progress, and Forecasting

In general in project reporting, project status refers to the state of the project at a given date; progress is the performance of the project since the last status date; and forecasting covers the expected project performance from this status date to the next status date and to the end of the project.

As explained above in the project tracking, the project manager collects project performance data periodically from the project team. This can be done by using activity reporting forms and during project meetings where accomplishments are discussed, and issues are communicated. During meetings, the project manager will review with the team the project's parameters, notably scope, schedule, costs, risks, etc. Discussions with each member will also focus on what is expected to be accomplished to the next status date and to the end of the project. This may entail reestimating activities and reassessing risks.

From the collected data, the project manager can then proceed to draft the project report to be presented to the project sponsor. The report should contain three sections: the project status—"where are we at"; the project progress—"what have we done since the last status"; and the project forecast—"what we expect to do until the next status and to the end of the project".

As the primary communication vehicle between the project manager and the project sponsor, the report should contain at a minimum the following information:

- Project status (at a given date):
  - Deliverables produced against planned
  - Current schedule to planned schedule
  - Current costs to planned costs

- EVM and variance analysis
- Risks register
- Change requests on hand
- Issues on hand
- Project progress (since last status):
  - Deliverables produced and accepted
  - Activities completed and in process
  - Costs incurred
  - Risk plans executed
  - Change requests applied
  - Resolved issues
  - Nonproductive time
- Project forecast (to next status)
  - EVM and trend analysis
  - Deliverables to produce
  - Activities to be completed
  - Costs to incur
  - Risk plans anticipated to be executed
  - Implementation of pending change requests
  - Resolution of pending issues
- Project forecast (to end of project)
  - EVM and trend analysis
  - Estimated final project delivery date and variance
  - Estimated final project cost and variance
  - Additional funding requirements, where necessary
- Anticipated risk events

The project report is the main presentation instrument to the project sponsor. Other reports may be required depending on the agreed communications plan. These could cover scope management and changes, specific risk evaluations, financial data, organizational and stakeholder readiness, contractor management, etc.

All project information collected and generated during project implementation is to be stored in an accessible central project repository. Many electronic systems exist on the market that the project manager can use. The information to be kept includes the following:

- Team member activity reporting forms
- Team member timesheets

- Equipment and materials usage reports
- Project performance on schedule and costs
- Project reports, including:
  - Change control
  - Issues
  - Deliverable acceptance
- Risk control reports
- Meeting minutes
- All correspondence

7.5 PROJECT EARNED VALUE MANAGEMENT

Earned value management (EVM) determines how much of a project has been completed at specific points in time. It is a performance variance measurement technique that compares the cumulative *value* of the work performed (earned) to both the *budgeted* cost of work scheduled (planned) and to the *actual* cost of work performed (actual). EVM utilizes monetary values to measure both financial and schedule performances of the project (examples in this section will use the $U.S. as the monetary currency).

EVM allows the project manager to monitor progress by measuring the performance of individual tasks and determine the overall performance of the project.

EVM provides no provision to measure project scope and quality, so it is possible for EVM to indicate a project is under budget and ahead of schedule, whilst its scope is either under-delivered or of low quality. EVM is only one tool in the project manager's toolbox.

7.5.1 Earned Value Management Terminology

EVM involves capturing primary key values for each schedule activity or work package—task—then determining derived data values. All values are based on the date the EVM measure is performed on the project. The following list of these values is explained further below:

7.5.1.1 EVM Primary Key Values

- **Budget at completion (BAC):** Total budgeted cost of the project.
- **Planned value (PV):** Cumulative budgeted cost for the task, scheduled to be completed up to the date of performance measurement.

- **Actual cost (AC):** Cumulative cost incurred for the work accomplished on the task up to the date of performance measurement.
- **Earned value (EV):** EV is the value for the work completed on the task up to the date of performance measurement. Describes the percentage completion of the planned value.

7.5.1.2 EVM Derived Data Values

- **Cost variance (CV):** The cost variance performance of the task, and the project, derived as the difference between the earned value (EV) and the actual cost (AC)
- **Schedule variance (SV):** The schedule variance performance of the task, and the project, derived as the difference between the earned value (EV) and the planned value (PV)

The EVM terminology also includes the following (explained further below):

- Cost performance index (CPI)
- Schedule performance index (SPI)
- Estimate at completion (EAC)
- Estimate to complete (ETC)
- Variance at completion (VAC)

7.5.2 Collecting Actual Costs Incurred

The project must institute a timely collection of the costs incurred for the work performed. The actual cost is posted against the respective task.

Certain difficulties lie in the cost collection at task level and are dependent on the organization's mechanism for actual cost recognition. First, the actual cost must correspond in definition and coverage to the planned cost (e.g., labor direct hours only, direct costs only, or all costs including equipment, materials and indirect costs). Second, the actual costs must be related to a task. This can prove difficult when actual costs of the project are provided by the financial department and cover many tasks. Third, the frequency of cost collection is to be compatible to the project manager's responsiveness needs. Monthly actual cost reporting is useless for projects that have tasks of an average duration in days, as the project manager would be ineffective in measuring appropriately earned value performance.

7.5.3 Determining Task Completion and Earned Value

Determining task completion is the key parameter to establish the earned value against the planned value.

The project manager and the team members need to agree on an acceptable, meaningful, and as realistic as possible way to determine the completion status of a task. This is often given as a percentage of completion. For effort-driven linear productivity work, for example, painting a wall, digging a trench, photocopying plans, etc., this does not present too much of a problem as the measure of progress is tangible. Difficulty lies when determining progress for nonlinear productivity knowledge-based work. In these cases, an estimated percentage complete is based on the individual's appreciation of progress. Care should be taken not to fall in the "90 percent syndrome"—when the task is considered 90 percent complete, but there is still 90 percent more work required to accomplish it.

A common technique is for the team to define "progress rules." The simplest method is to apply a 0/100 rule, where no progress is posted for the task until it is finished. Another rule is the 50/50 rule, where 50 percent progress is posted when the task is started, and the remaining 50 percent is posted upon completion. Other progress rules such as a 25/75 rule or 20/80 rule assign more weight to the latter stage of completing. These progress rules are well adapted for projects with planned task durations of less than two weeks of elapsed time.

Depending on the task progress, the earned value is calculated as follows:

$$EV = \% \text{ complete of PV}$$

- Examples for a task with a PV of $10,000:
  - The task has been 100% completed: EV = $10,000
  - The task has not started: EV = $0
  - The task has started and is 65% complete: EV = $6,500
- The task has started and the 20/80 is applied: EV = $2,000

In the example in Figure 7.15, the project implementation start date is March 1, and the earned value report date is April 20. Tasks A through D have been completed, while E and F have been started but as yet are not finished. The BAC column indicates the planned value of each task and the total for the project. The "PV to this date" is the planned value at April 20. The AC column presents the collected actual costs incurred The EV can be calculated for each task, as shown.

Earned Value at today's date = 20-4-yyyy

Task	Planned Start Date	Planned End Date	BAC	PV to this Date	AC	% Comp.	EV	SV	SPI	CV	CPI
A	1-3-yyyy	13-3-yyyy	10,000	10,000	10,500	100	10,000	0	1.00	−500	0.95
B	1-3-yyyy	19-3-yyyy	8,500	8,500	7,500	100	8,500	0	1.00	1,000	1.13
C	14-3-yyyy	6-4-yyyy	15,000	15,000	14,500	100	15,000	0	1.00	500	1.03
D	20-3-yyyy	4-4-yyyy	12,000	12,000	11,000	100	12,000	0	1.00	1,000	1.09
E	4-4-yyyy	25-4-yyyy	35,000	30,000	47,000	40	12,000	−18,000	0.40	−35,000	0.26
F	6-4-yyyy	25-4-yyyy	47,000	40,000	8,000	20	8,000	−32,000	0.20	0	1.00
G	25-4-yyyy	9-5-yyyy	28,000	0	0	0	0	0	0	0	
H	10-5-yyyy	31-5-yyyy	16,000	0	0	0	0	0	0	0	
	Total Project	Budget at Completion	128,000								
		Today's Date		115,500	98,500		65,500	−50,000	0.57	−33,000	0.66

FIGURE 7.15
Earned value example

7.5.4 Calculating Cost and Schedule Variances

Cost and schedule variances are calculated at each task level and aggregated for the project at the earned value report date.

- **Cost variance (CV):** The cost variance performance of the task, and the project, derived as the difference between the earned value (EV) and the actual cost (AC)

$$CV = EV - AC$$

- **Schedule variance (SV):** The schedule variance performance of the task, and the project, derived as the difference between the earned value (EV) and the planned value (PV)

$$SV = EV - PV$$

- **Budget at Completion (BAC):** The planned cost of the project
- This can plotted and illustrated as shown in Figure 7.16.

Proceeding from the calculated example given here (Figure 7.17), we arrive at :

- **Estimate at Completion (EAC):** The revised planned cost of the project

$$BAC / CPI$$

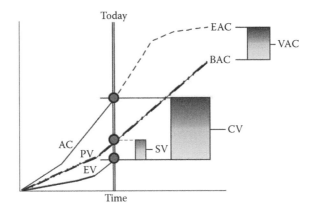

FIGURE 7.16
Earned value graphic

- **Variance at Completion (VAC):** The difference to the planned cost

$$BAC - EAC$$

- **Variance At Completion (VAC):** Variance of total cost of the work and expected cost.

$$VAC = BAC - EAC.$$

7.5.5 Trend Analysis and Forecasting

From the captured and calculated values, a basic trend analysis can be performed that can be used for forecasting. The cost and schedule variances

Earned Value at today's date = 20-4-yyyy

Task	Planned Start Date	Planned End Date	BAC	PV to this Date	AC	% Comp.	EV	SV	SPI	CV	CPI
A	1-3-yyyy	13-3-yyyy	12,000	12,000	11,000	100					
B	1-3-yyyy	19-3-yyyy	10,000	10,000	10,500	100					
C	14-3-yyyy	6-4-yyyy	20,000	20,000	22,000	100					
D	20-3-yyyy	4-4-yyyy	15,000	15,000	13,000	100					
E	4-4-yyyy	25-4-yyyy	22,000	18,000	24,000	75					
F	6-4-yyyy	25-4-yyyy	30,000	28,000	25,000	80					
G	25-4-yyyy	9-5-yyyy	25,000	0	0	0					
H	10-5-yyyy	31-5-yyyy	15,000	0	0	0					
	Total Project	Budget at Completion	149,000								
		Today's Date		103,000	105,500						

FIGURE 7.17
Earned value worked example

are converted into performance efficiency ratios as follows:

- **Cost performance index (CPI):** The cost performance efficiency of the task, and the project, derived as the ratio between the earned value (EV) and the actual cost (AC)

$$CPI = EV / AC$$

- A CPI value less than 1.0 indicates a cost over-budget
  - A CPI value greater than 1.0 indicates a cost under-budget

- **Schedule performance index (SPI):** The schedule performance efficiency of the task, and the project, derived as the ratio between the earned value (EV) and the planned value (PV)

$$SPI = EV / PV$$

- A SPI value less than 1.0 indicates behind schedule
- A SPI value greater than 1.0 indicates ahead of schedule

The project manager would have established with the project sponsor the acceptable +/− range of the CPI and SPI that can be under the control of the project manager. Indices outside of the range would require escalation to the project sponsor. The most common range is +/− 10 percent, which results in a range of 0.90 to 1.10, within which the project manager can take variance corrective measures without referral to the project sponsor.

From the variance analysis and the efficiency performance ratios, the project manager and the team can derive the following:

- **Estimate to complete (ETC):** The expected cost required to complete the project
- **Estimate at completion (EAC):** The expected final total cost of the project
- **Variance at completion (VAC):** The expected cost variation at the end of the project

The ETC can be calculated in three ways, depending on analysis and reasons for the variances:

- **ETC based on new estimate:** When the variance analysis demonstrates that the original basis of estimate is wide of the mark and that the trend analysis cannot be used as a reference, the project manager and the team proceed to develop revised estimates for the remaining work to perform.

$$ETC = \text{New estimate to complete remaining tasks}$$

- **ETC based on typical variances:** When the variance analysis demonstrates that the current trend is considered as typical of future variances, the project manager can then use the index to extrapolate the estimate for the remaining work to perform.

$$ETC = [(BAC - EV)/CPI]$$

- **ETC based on atypical variances:** When the variance analysis is considered as atypical, due to unusual conditions that are not expected to be repeated, the project manager can consider that the current estimates are acceptable for the remaining work to perform.

$$ETC = (BAC - EV)$$

The EAC can then be derived as follows:

- EAC based on new estimate:

$$EAC = AC + ETC$$

- EAC based on typical variances:

$$EAC = BAC / CPI$$

- EAC based on atypical variances:

$$EAC = AC + (BAC - EV)$$

Finally, VAC can be established as:

$$VAC = BAC - EAC$$

7.5.6 Exercise: Earned Value Reporting

From the earned value report below, complete the columns EV, SV, SPI, CV, and CPI.

Next calculate the EAC based on a current trend considered as typical of future variances (see Figure 7.18).

Earned Value at today's date = 20-4-yyyy

Task	Planned Start Date	Planned End Date	BAC	PV to this Date	AC	% Comp.	EV	SV	SPI	CV	CPI
A	1-3-yyyy	13-3-yyyy	12,000	12,000	11,000	100					
B	1-3-yyyy	19-3-yyyy	10,000	10,000	10,500	100					
C	14-3-yyyy	6-4-yyyy	20,000	20,000	22,000	100					
D	20-3-yyyy	4-4-yyyy	15,000	15,000	13,000	100					
E	4-4-yyyy	25-4-yyyy	22,000	18,000	24,000	75					
F	6-4-yyyy	25-4-yyyy	30,000	28,000	25,000	80					
G	25-4-yyyy	9-5-yyyy	25,000	0	0	0					
H	10-5-yyyy	31-5-yyyy	15,000	0	0	0					
	Total Project	Budget at Completion	149,000								
		Today's Date		103,000	105,000						

FIGURE 7.18
Earned value exercise

7.6 PROJECT DOCUMENTATION MANAGEMENT

All projects involve the production and control of documentation. There will be many types of documents with varying purposes, natures, and life cycles. Information technology is commonly used to share and convey documentation with Web-enabled software.

The documentation management system provides an efficient way of sharing knowledge and information among the project's participants. All participants should have easy access to the project's documentation repository to lodge documented information, and to consult and search for content that is relevant to their interests.

Project documentation management and control is closely related to configuration management, which describes the technical and administrative control of the deliverable components.

7.6.1 Project Documentation Planning Overview

Project documentation planning establishes the processes for the development, maintenance, and retirement of the documentation. Project documentation planning identifies and describes:

- What is managed, i.e., the system documentation under configuration control
- How to manage this controlled documentation
- Documentation management roles and responsibilities
- Key documentation milestone schedule

The project documentation processes define the management of the development, packaging, delivery, and maintenance of documentation. This includes documentation change control for a formal control of information development, dissemination, changes, and retirement.

Furthermore, the project documentation plan provides management with visibility on documentation status and cost.

7.6.1.1 Types of Documents to Manage

Each project will manage and handle a variety of documents that are specific to the project and to the product/service to deliver. Some documents will evolve during project implementation such as the project scope of work, whilst others will be created during the phase. The list below offers a wide range of documents that can be used, and to which other documents can be added:

Acceptance plan	Product descriptions
Change request management plan	Project cost plan
Change request register	Project plan
Communications management plan	Project progress reports
Contractor management plan	Project schedule
End of project report	Project scope of work
Exception reports	Quality management plan

Financial reports	Resource management plan
Follow-on recommendations	Risk management plan
Formal acceptance statement	Risk register
Installation plan	Stage plans
Issue register	System test plan
Lessons learned report	Training plan
Meeting minutes	Transition plan
Post implementation review	User acceptance testing results
Procurement management plan	User testing plans

7.6.1.2 *Project Documentation Categories*

Project documents can be generally grouped under three categories:

- Permanent documentation:
  - Project deliverables (contractual, acceptance results reports, user manuals, training materials, etc.)
  - Delivered product/service support and maintenance (contracts, design specifications/drawings, database definitions, process diagrams, etc.)
- Temporary documentation:
  - Internal project deliverables no longer needed once the project has been completed (draft documents, interim progress reports, etc.)
- Internal communication (meeting minutes, working papers, etc.)

All permanent documents will form part of the official project hand-over and require strict control management. Temporary documents require the same control and may need to be archived at the end of the project, depending on the discretion of the project manager or the performing organization's needs.

7.6.2 Documentation Management Functions

Project documentation management should cover the following major functions:

- Catalogue of all documents and deliverables under change control
- Key information per document:
  - Description
  - Purpose/Objective

- Form and format
- Responsibilities for:
  - Production
  - Review
  - Approval
- Retention and usage (temporary or permanent, internal or project deliverable)
- Process for quality assurance
- Status information per document:
  - Planned date of completion
  - Current status and effective date
  - Projected date of completion
  - Change history
- Management of multiple versions
  - Version control
  - History management
- Secure storage of documents
- Authorization for:
  - "Checking out" a copy of a document for update
  - "Checking in" an updated version of a document
- Accessibility:
  - Search and viewing authorization
  - Viewing and consulting documents, present and past versions
- Support Internet, intranet, or WAN networks

7.6.3 Maintaining the Documentation Database

Present technology provides for electronic means to manage project documentation. The project manager should institute a structured project documentation format and naming convention, which will provide:

- A central repository of all documents with cross-references
- A consistent approach to documents of equal formats and tables of contents
- A common nomenclature for project documents
- Date and version control
- Revision history

7.6.4 Using Electronic and Web-enabled Technologies

Web-based software provides project managers, team members, project sponsor, stakeholders, suppliers, and clients to access, view, share, and edit documents, and manage information easily and across a wide geography.

When choosing a project document management system, the project manager should ensure that the functionalities provide for storage, indexing and retrieval, versioning, and security. Most systems for managing documents address the following areas:

- Creation: the process to create a new document
- Storage: the type of repository support and its capacity
- Versioning: the method for version control and history logging
- Indexing: the method used to organize and index documents
- Retrieval: the browsing and searching capabilities
- Retention and archiving: the method used for long-duration storage
- Traceability: the method for identifying usage and cross-referencing documents
- Security: the process for authorized access and disaster recovery

7.6.5 Exercise: Project Documentation

Review your current project and list all the key documents that are currently under change control and describe the process.

7.7 PROJECT ACCEPTANCE

The purpose of obtaining project acceptance is to formally recognize that all deliverables produced during the project implementation phase have been completed, tested, accepted, and approved by the project's client/customer and sanctioned by the project sponsor (see Figure 7.19).

The acceptance signifies that the product or service has been successfully transitioned to the performing organization. Formal acceptance and approval also indicate that the project is ready for project closeout.

7.7.1 Obtaining Acceptance Signature

As the deliverables of the project are produced and accepted, approval signatures are obtained. These may come from the project sponsor, key stakeholders and/or client/customer decision-makers. By obtaining all acceptance approvals and signatures, the project manager can then obtain a final acceptance declaration signed and approved by the sponsor stating that the project is complete and ready to proceed to the project closeout phase (see Figure 7.20).

Project Acceptance Checklist		
Project Id.	**Project Description**	**Date**

Deliverable	Reference	Comments	Date	Initials

FIGURE 7.19
Project acceptance checklist

Project Acceptance Declaration

Project Name:		Pj Id.		Version		Date	
Customer/Client:							
Sponsors & Owners:							
Project Manager:							

I (we), the undersigned, acknowledge and accept delivery of the
above mentioned product/service on behalf of .... Company.
My (our) signature(s) attest to my (our) approval of the application
and my (our) acceptance of this product/service this date ....
Without exception, I (we) believe that this product/service meets
the requirements and needs of .... Company that were
set forth at the initiation of this project. I (we) understand that the
.... Company can now exploit/use the product/service with
the provisions and remarks set below, and that any future
changes, modifications, enhancements or improvements, will be
provided by the support group.

I (we) and (are) the duly appointed representative (s) of the ....
Company.

For The .... Company:
Provisions and remarks:
Signature:
Date:

FIGURE 7.20
Project acceptance declaration

7.7.2 Conduct Final Status Meeting

Once the project has been successfully approved and transitioned to the
performing organization, the project manager should prepare a final status report and conduct a final status meeting. These will be used, along
with the cumulated project documentation, to complete the project and
project management life-cycle phases. Any outstanding or pending action
items must be recorded along with a plan to address and resolve them.

8

Project Closeout

8.1 CHAPTER OVERVIEW

The project Closeout phase chapter explains the process and steps to follow for the final phase of the project before formal handover and completion. Focus is placed on the activities to perform to prepare the post-implementation review and the project closeout reports.

The project approval, administrative and contract closures, as well as the collection of lessons learned, are described, with special emphasis on the organizational readiness needs for the performing organization.

Team members' performance evaluation and recognition complete this module.

8.2 PROJECT CLOSEOUT OVERVIEW

Project closeout is the final phase of the project development life cycle. The closeout of a project is just as important as the other phases and the project manager and sponsor must allocate the time and effort to thoroughly complete this final phase.

The Project Management Institute (PMI) defines project closeout and evaluation as "formalizing acceptance of the project or phase and bringing it to an orderly end."

The purpose of project closeout is to capture key project metrics, assess and evaluate the project, perform all applicable administrative actions,

Project Close-Out Overview

Administrative Closure

Organizational Readiness

People Management

Contract Closure

Project Closure & Approval

Lessons Learned

FIGURE 8.1
Project closeout overview

provide performance feedback on project manager and team members, update the skills inventory when and where needed, derive any lessons learned and best practices to be applied to future projects, and collate and archive project documents into the project repository (see Figure 8.1).

8.2.1 Planning Project Closeout Activities

Project closeout activities are identified, scheduled, and their cost estimated at the project planning phase. These activities are reviewed and modified throughout the implementation phase as a result of project work performance. The project closeout steps include:

- Producing the post-implementation review report
- Accepting the project's product/service by obtaining sponsor and customer/client sign-off
- Conducting administrative and contractual closures
- Demobilizing resources—personnel, equipment, and facilities
- Team performance evaluation and recognition of outstanding achievements
- Conducting a lessons-learned session
- Completing and archiving project records

FIGURE 8.2
Project closeout reports

- Celebrating project completion
- Project official handover to the performing organization
- Producing the project closeout report

The two major outputs of the closeout phase are shown in Figure 8.2:

- Post-implementation review report
- Project closeout report

8.2.2 Post-Implementation Review Report

Project closeout begins with a post-implementation review to collate project metrics and evaluate total project performance. The information collected and evaluated is subsequently used to produce the post-implementation review report.

The post-implementation review is conducted with team members, customers, and other stakeholders to examine and verify project scope delivery compared to the original project scope. The review also covers comparisons of actual performance to planned schedule, budget, quality, and resource utilization. Change requests implemented are appraised and the applied risk plans are assessed for their effectiveness. Team performance is evaluated and stakeholder relationship is gauged relative to communication, project control, project reporting, conflict management and resolution, cross-functional cooperation and accountability, roles and responsibilities, and completion and success criteria. Best practices and lessons learned are recorded (see Section 8.8 below).

The project manager should consider inviting project managers from the performing organization with experience on similar projects and prior knowledge, who can provide information and insight on the review process.

8.2.2.1 Key Project Metrics

The project manager collects and reviews the set of project progress and status reports for the duration of the implementation phase, and summarizes the results (see Figure 8.3).

The number of metrics categories collated will depend on management needs, and should include as a minimum the following:

- Scope
  - Number of baseline deliverables
  - Number of deliverables achieved at project completion
- Schedule
  - Number of milestones in baseline schedule
  - Number of baseline milestones delivered on time (according to final baseline schedule)

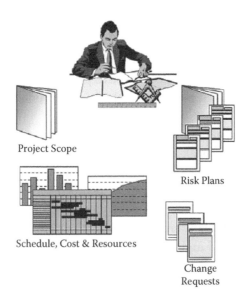

Project Scope

Risk Plans

Schedule, Cost & Resources

Change Requests

FIGURE 8.3
Project metrics

- Variance in elapsed time of original schedule and final actual schedule
  - Variance in elapsed time of final baseline and final actual schedule
- Cost
  - Variance between the final cost, final approved baseline cost estimate, and the original cost estimate
  - Number of performed budget baseline updated estimates
- Change requests
  - Number of approved scope changes introduced in the implementation phase
  - Number of changes rejected or suspended
  - Schedule and cost variations originating from applied approved changes
- Risks
  - Number of risk plans identified and performed
    - Mitigated threats and opportunities
    - Contingent threats and opportunities
  - Number of risk plans identified and abandoned
  - Number of new mitigation risk plans introduced and performed
  - Number on contingency plans identified
- Contingency reserve utilized

8.2.2.2 Post-Implementation Review Questions

The post-implementation review is an integrated part of the ongoing implementation phase and can be performed at the same frequency as the scheduled project progress and status review meetings. At project closeout, the project manager performs a final review and adjoins it to all previous reviews. Special focus is given to review meetings with representatives of the operational performing organization.

A questionnaire is established to capture comments and remarks from all those concerned and/affected by the project delivery. Questions can be open, with the challenge to convert what could be subjective comments to objective measures, or scaled from low to high; dissatisfied to very satisfied, etc. (See Figure 8.4.)

Below is a sample of typical post-implementation review questions:

- Delivery effectiveness:
  - Overall assessment of the outcome of this project

FIGURE 8.4
Review questions

- Adherence of the product/service to the needs of the performing organization
- Extent the business case objectives and goals were met
- Scope, schedule, and cost management:
  - Scope variance to initial project baseline
  - Project schedule baseline
  - Actual cost expenditure variance to initial project budget
  - Effectiveness of change control process for scope, schedule, and cost
  - Effectiveness of the quality management plan
- Transition management:
  - Effectiveness of efforts to prepare the organization for the transition of the product/service of the project
  - Effectiveness and timeliness of training
  - Effectiveness and completeness of the project product/service documentation
  - Effectiveness of the support received during product/service implementation
- Risk and issue management:
  - Effectiveness of risk management process
  - Effectiveness of issues management process
- Communications management:
  - Effectiveness of communications plan and process
  - Effectiveness of information management:
    - Collection and distribution
    - Accessibility
  - Effectiveness of information sharing:
    - Meetings and minutes
    - Reporting
    - Documentation completeness

- Acceptance management:
  - Effectiveness of acceptance management process
  - Completeness of project deliverables
  - Completeness of supporting documentation
- Performance evaluation:
  - The performing organization:
    - Sponsorship and stakeholder involvement
    - Operational involvement
  - The project team:
    - Overall project manager performance
    - Effectiveness of project team roles and responsibilities
- Individual project team member performance

8.2.3 Project Closeout Report

The project closeout report (see Figure 8.5) is prepared by the project manager and distributed to the project sponsor and key stakeholders. An end of project review meeting is held to present the main aspects of the closeout report and lessons learned.

The content of the report is specific to each project; however, it is often a summary of the project metrics, the post-implementation review report and the outcomes of the different closures. A key section of the project closeout report concerns the lessons learned.

8.2.4 Exercise: Project Closeout Review

> Review your current project and draft a questionnaire for a closeout review, identify the target audience, and establish an interview schedule.

8.3 PROJECT CLOSURE AND APPROVAL

The project is considered complete when it has been successfully implemented and transitioned to the performing organization and approved by the project sponsor and customer/client. At this point in the project

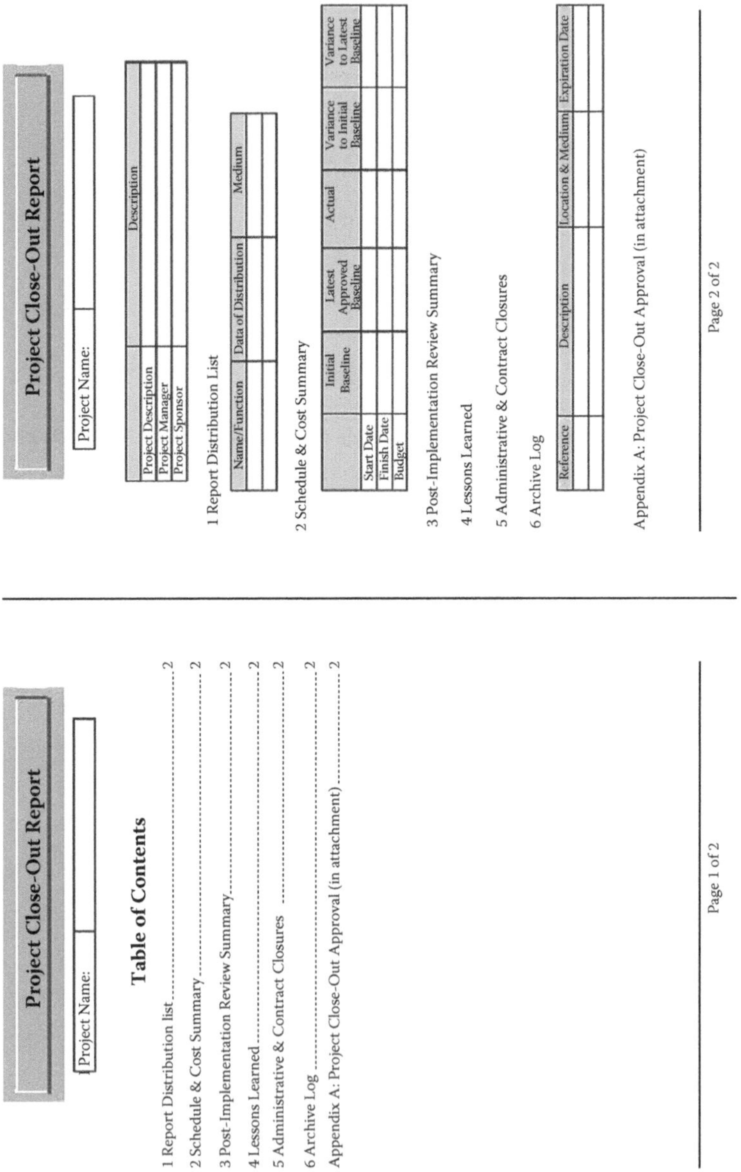

Project Close-Out Report

Project Name:

Table of Contents

Page 1 of 2

Project Close-Out Report

Project Name:

	Description
Project Description	
Project Manager	
Project Sponsor	

1 Report Distribution List

Name/Function	Data of Distribution	Medium

2 Schedule & Cost Summary

	Initial Baseline	Latest Approved Baseline	Actual	Variance to Initial Baseline	Variance to Latest Baseline
Start Date					
Finish Date					
Budget					

3 Post-Implementation Review Summary

4 Lessons Learned

5 Administrative & Contract Closures

6 Archive Log

Reference	Description	Location & Medium	Expiration Date

Appendix A: Project Close-Out Approval (in attachment)

Page 2 of 2

FIGURE 8.5
Project closeout report template

FIGURE 8.6
Project closure and approval

development life cycle, the responsibilities of the project manager are to assess how closely the project met customer/client needs. The activities performed to produce the post-implementation review provide the information required to meet those responsibilities (see Figure 8.6).

8.3.1 Scope Verification

This is the process of obtaining the stakeholders' formal acceptance of the completed project scope and associated deliverables. Verifying the project scope includes reviewing deliverables to ensure that each is completed satisfactorily. If the project is terminated early, the project scope verification process should establish and document the level and extent of completion.

8.3.2 Meeting Project Goals and Deliverables

The most important measures of the success of a project are whether the product/service was developed and delivered successfully and how well the needs have been met.

Depending on the size and type of the project and the structure of the performing organization, the project manager gathers feedback using different surveys for different stakeholder groups. The surveys collect sufficient information for assessing the success of the project in meeting its goals and the effectiveness of the deliverables as well as the performance of the project team and performing organization. The collected survey will be used at formal project acceptance and closure.

8.3.3 Ensuring Project Delivery to Scope and Approved Changes

The project managers and the core team members compare the delivered scope to the original approved baseline scope, plus all incorporated changes as planned and implemented, and substantiate and explain variances to the project sponsor and customer/client.

Reference is made to the risk register for applied risk plans, and to the change request register to demonstrate that changes have been approved.

8.3.4 Completion of All Project Activities

The project manager and the core team members proceed to compare the actual work performed to the work defined in the original WBS, plus the additional scheduled tasks resulting from the project implementation. Any activity pending is reported and scheduled to be completed after project handover.

8.3.5 Commissioning and Approvals

Where necessary, all commissioning and certifications for proper and legal use of the delivered product/service are collected. Prior to final payment, the contractor is to provide the necessary health, safety, and environmental (HSE) certifications issued by the formal authorities.

8.3.6 Securing Project Acceptance

The project manager will review the success measurement and completion status with the project sponsor and customer/client (see Figure 8.7). Formal acceptance and closure will indicate the completion of the project. Where necessary, partial acceptance may be given when agreement is reached,

Project Close-Out Approval

	Description
Project Name	
Project Description	
Project Manager	
Project Sponsor	

The undersigned acknowledge they have reviewed the Project xxxxxx and agree with the approach it presents.
[*List the individuals whose signatures are required. Examples of such individuals are Project Sponsor, Business Owner, Performing Organization Representative ane Project Manager, Add additional signature lines as necessary.*]

Signature:	Date:
Print Name:	
Title:	
Role:	

Signature:	Date:
Print Name:	
Title:	
Role:	

Signature:	Date:
Print Name:	
Title:	
Role:	

FIGURE 8.7
Project closeout approval form

that outstanding issues do not halt the start-up of the operation. These issues, however, will need to be resolved and corrected within an agreed timetable.

8.3.7 Exercise: Project Closure and Approval

Review your current project and assess how the project closure has been conducted, or if yet to be done, how you would plan for this.

8.4 ADMINISTRATIVE CLOSURE AND FINANCIAL REPORTING

The purpose of administrative closeout is to perform all administrative tasks required to bring the project to an official close (see Figure 8.8). These tasks cover the preparation of the required reports as identified in the communications plan, closing contracts (see Section 8.5), evaluation the performance of the team, stakeholders and the performing organization (see Section 8.7), and proceeding to archiving the project documentation.

Immobilized equipment and other infrastructure items used in the project's inventory are accounted for and returned to the proper owner. Moreover, because many projects also require that facilities be rented and provisioned, it is the project manager's responsibility to ensure that all facilities and any related equipment are properly distributed or closed out.

8.4.1 Internal Status Reports

Based on the project closeout report, the project manager prepares for internal usage to stakeholders and other managers of the performing

FIGURE 8.8
Administrative closure

organization, the set of reports as identified in the communications plan. These would include such topics as: scope, schedule and cost variances, resource usage, installation and testing, and training results, etc.

8.4.2 Internal Financial Reports

Each organization would have its specific needs for project financial reporting, and the project manager will also need to provide particular formats to be aligned to the financial department's requirements. Irrespective of the situation, project financial reports will primarily concern with reporting on the direct costs of the project and whatever indirect costs are affected to it.

The core of most reports will consist of the following categories and present the total accrued actual costs compared to the original and latest approved budget.

- Human resources—internal personnel
  - Actual costs accrued by staff within the project organization who are directly involved in the project. This category does not include contractual staff.
- Materials and consumables
  - This category includes tangible items with a per-unit cost valued as direct costs to the project.
- Equipment (procured, with depreciation, rented, or leased)
  - Equipment: Only equipment items valued as direct costs to the project.
- Travel and expenses
  - Including accommodation, meals, and subsistence allowances.
- Contractual
  - Costs for subcontracted work completed external organizations for the project organization. This includes design and engineering services, products provided that are not part of the equipment category.
- External expertise and services
  - Costs for fees to external consultants, such as legal advisors, independent auditors, etc.
- Indirect cost breakdown
- Costs referring to overheads attributed to the project.

8.4.3 Internal Human Resources (HR) and Other Shared Services Reports

These reports would have been identified in the communications plan and would follow formats provided by the destination departments.

Specific reports to HR would include team member performance evaluations as well as any skills and competences gained during the project.

8.4.4 Archiving

Throughout the course of the project, and specifically during the implementation phase, the project manager maintains a project repository to store the project documentation data. As the project progresses, the repository serves as a central point of reference for all project documents and provides an audit trail recording the history and evolution of the project at its official closure.

During project closeout, the project manager should examine the repository to ensure that it contains the total and latest set of the produced and relevant project-related documents. In addition, both the post-implementation review report and project closeout report are included. The project repository becomes the project archive and is stored electronically. The archive should include a description of each document, the application (including version) used to create the archived materials, and a point of contact if further information is needed.

Additionally a predefined selection of documents is held in hard-copy format and archived in a designated documentation area, following corporate retention guidelines.

A project archive would typically contain the following:

- Project supporting documentation, including the business case and project proposal
- Project description/definition documents such as the project charter and project plan
- Project documents defining approved baseline scope, schedule, and cost
- Project schedules (all versions)
- Project financial reports
- Risk plans and risk register
- Issues log and details (open and resolved)
- Change request register

- Project status reports
- Team member progress reports
- Test, installation, and audit results
- Correspondence, including any pivotal or decision-making memos, letters, e-mail, etc.
- Meeting notes
- Project acceptance checklists by deliverable
- Project deliverable approval, with original signatures
- Final project acceptance, with original signatures
- Post-implementation review report
- Project closeout report

8.4.5 Exercise: Administrative Closure

Establish a list of closeout internal reports required for your performing organization, and structure the information contents of your documentation archiving system.

8.5 CONTRACT CLOSURE

Contract closure, as depicted in Figure 8.9, is concerned with completing and settling the terms of the contract (see Chapter 6, Procurement Management). It supports the project closeout process in determining if the performance described in the contract was completed accurately and satisfactorily. The project manager should be conversant with those contracts that have specific terms or conditions for completion and closeout. These clauses were established during the procurement steps.

Not all projects require the contract closure process, as this process applies only to those phases, deliverables, or portions of the project that were performed under contract.

The contract closure process provides formal notice to the contractor (provider), in written form, that the deliverables meet the project's expectations, are acceptable and satisfactory. This formalizes acceptance and

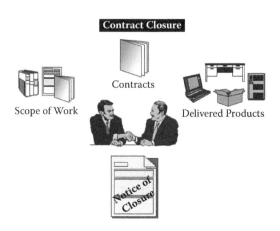

FIGURE 8.9
Contract closure

closure of the contract. The provisions for formal acceptance and closure are described in the contract itself.

If the product or service does not meet the expectations, the contractor will need to correct the problems before a formal acceptance notice is issued. The project manager should plan quality audits to be performed during the course of the project, giving the vendor the opportunity to make corrections as soon as the problem occurs. This avoids accumulating problems and issues on the provider to be resolved at project closure, which consequently delay total project delivery.

When a procurement department holds responsibility for contract administration, the project manager informs when the contract is deemed fulfilled and the department then follows the formal procedures to give notice to the provider that the contract is complete.

8.5.1 Approval of Contractor Deliverables

The deliverables prepared and submitted by the contractor are subject to review by a review committee, usually the project manager or a contracting authority, such as the procurement department, in order to ensure that they are in compliance with the predefined acceptance criteria as described in the terms and requirements of the contract. Reviewers are to be qualified and experienced in the specific subject.

In most cases, the project manager or a designated project member would act as the deliverable reviewers. The project manager oversees the process as follows:

- On receiving the deliverable by the contractor, the reviewers examine whether the deliverable responds to the contract requirements, according to the acceptance criteria.
- Results/conclusions of the review are drafted, stating either final approval, or conditional approval or rejection of the specific deliverable.
- The decision is notified in writing to the contractor by the project manager, at the same time presenting the reasons in case that it rejects the deliverable or requires further modifications to them. In many cases, a partial acceptance is given (see Section 8.5.2). The project manager is required to inform the contractor on the decision within a time limit as specified in the general conditions of the contract. Failure to comply with the established predefined timeframe would usually explicitly approve the deliverables.

8.5.2 Managing "Punch Lists," "Snag Lists," and Warranty for Handover

Partial acceptance of the contractor deliverables can be approved when the project manager and the contractor mutually agree that a substantial portion of the contract has been satisfactorily completed.

Both parties agree on a list of items to be completed or corrected (i.e., punch list or snag list) and establish the schedule for the contractor to complete the items. Partial acceptance will allow for partial payment to the contractor. The delay for the correction of the punch-list items is often included in the product/service warranty period.

8.6 ORGANIZATIONAL READINESS

Organizational readiness concerns the activities that ensure the performing organization's capability to operate under the new situation to be delivered (see Figure 8.10). The project manager must coordinate these efforts with the performing organization's change management plan.

FIGURE 8.10
Organizational readiness

Preparing for organizational readiness in the project closeout phase includes:

- Handover to operations
- Education and training needs
- Operational ramp-up
- Support issues

8.6.1 Handover to Operations

Prior to handover to the performing organization, the project manager ensures that all facilities are ready and operational, such as securing physical space, installing appropriate equipment, obtaining the appropriate building permits, etc.

The necessary documentation provided with the delivered product/service is produced and distributed. This will include:

- User manuals
- Online help
- Assembly or usage instructions

8.6.2 Education and Training Needs

The planned training schedule is completed and evaluated to ensure that all operational personnel have been appropriately trained in the use of the delivered product/service.

8.6.3 Operational Ramp-Up

The project manager is responsible for transition into the existing operational environment, according to a pre-established philosophy: from a phased, stepped transition, as for a progressive office move, to a one-step "switchover," as for an overnight change from an old to a new processing system.

8.6.4 Support Issues and Guarantees/Warranties

This ensures the supportability of the product/service in readiness for its operational use. This covers from the institution of internal "help" lines to specific support from external organizations with appropriate service-level agreements.

When the product/service support is to be performed by an internal department of the performing organization, maintenance manuals must be provided by the providers or manufacturers. The manuals or catalogue are to contain descriptive information and maintenance instructions; parts lists; usage instructions; names, addresses, and telephone numbers where replacement parts and service can be obtained; and all other information required for the performing organization to use, maintain, and service the product properly.

All applicable guarantees, warranties, and bonds for all materials and equipment incorporated into the delivered product/service are to be compiled by the project manager, and include written warrantees from contractor and subcontractors for the specified general warranty period.

8.6.5 Exercise: Organizational Readiness

Assess your current or previous project and review how the transition and handover steps were effective to meet the organization's operational needs.

Recognitions &
Evaluations

FIGURE 8.11
Performance evaluation

8.7 PEOPLE MANAGEMENT

This section will concentrate on the project closeout steps for people management. The topic is expanded in Chapter 9, Project Leadership Skills.

The project manager will primarily focus on all aspects affecting project team members as they complete their assignments and/or are present at the completion of the project (see Figure 8.11). All team members who have performed on the project will wish to be recognized for the work done. They will also wish to know how they fared and receive an evaluation of their performance. This evaluation can then be used by their respective line managers for a formal and objective position performance evaluation.

It is of great importance that any member of the project who has performed in any way be objectively recognized and evaluated, as this has a direct motivational effect on the individual's next assignment. Lack of recognition for good work done can often lead to despondency and below-par quality for the next assignment.

8.7.1 Recognition and Awards

It is important for the project manager to recognize the work and efforts of all team members. When success in a project is achieved, the project manager must not fail to provide recognition to the team. If individuals can be singled out for significant achievements, team recognition must also be done in a balanced way.

At project implementation start-up, the project manager sets and agrees with the team members, the success criteria that will be used to measure and recognize performance. Success should not be limited to the achievement of schedule and budget, and should cover the individual's skills enhancement in both hard skills (for example, learning a new technique) and soft skills (for example, the relationship with stakeholders).

The project manager may also express recognition of a successful team effort by praising the team at a key meeting or in the presence of the project sponsor and key stakeholders. Many individuals take pride not only in the work they perform but also that senior management is aware of the formal appreciation. Other formal recognition can be achieved through articles in the internal company newsletter that is circulated to the organization, and/or when appropriate in industry-specific publications.

The most powerful and human form of recognition, however, especially for knowledge workers, is informal. Even though this is not a tangible reward, a "thank you" is extremely potent. Other recognitions cover asking a member to lead a meeting, produce a document, or represent the project at an external meeting.

In general, the project manager should not wait until the end of the project to recognize and award. The project manager requests day-to-day performance and should provide day-to-day recognition. On a day-to-day basis, recognition will often be informal.

8.7.2 Performance Evaluation

The project manager and/or team leader document their feedback on the performance and accomplishments of each project team member. Feedback documentation, providing an honest and accurate evaluation, should be prepared and reviewed with the individual team members. Following mutual agreement on the validity of the performance evaluation, this is forwarded to each project team member's manager, to be used as input to the company performance appraisal process.

8.7.3 Celebrations

The project manager should officially recognize the project team's efforts and accomplishments and should celebrate and thank them for their participation and officially close the project. The celebration is to be appropriate to the type and size of the project and should be an event that formally recognizes the project end and brings closure to the team members' work.

Whatever type of celebration, the project manager should have planned the event and established a budget for its cost, as negotiated with the sponsor during the planning phase.

For long duration projects, the project manager should plan many celebration meetings to cater for those team members who complete their assignments at different times during the project implementation.

8.7.4 Preparing Team Members' Next Assignment

As team members reach the end of their assignment and approach their release date from the project, they will be concerned with their next assignments and/or their return to their original functional departments. The project manager should liaise with either the next project manager or the functional department manager for a suitable transition. This must cover the completion of the present assignment whilst not hampering the start of the next.

Releasing project team members is rarely an official process. However, it should be noted that at the conclusion of the project or work assignment period, project team members are released and return to their functional managers or are assigned to a new project. The project manager informs line managers, or other project managers, on the impending date of resource release. Depending on the project, this advice on resource release can intervene at least on a monthly basis.

Care should be taken by the project manager when a team member starts a next assignment while the present one is yet not completed. Agreement should be reached between all parties to allow the necessary time to the team member to accomplish the current task, as the next assignment begins, which will demand accrued focus and interest from the individual.

Reassigning personnel is not always an option for the project manager. Most often, the team members and all those who have supported the project work actually report to a functional manager and not directly to the project manager. Consequently, the project manager may not have the responsibility or authority to reassign personnel.

8.7.5 Disbanding the Core Team

Eventually, the project manager and the core team are the remaining individuals who have closed the project in an orderly manner. It remains now for the project manager to officially disband the project team. The release of the core team members would follow the same approach as that explained above.

8.7.6 Exercise: People Management

Describe the recognition and award mechanism used on your project, and define recommendations to make to your management to enhance this.

8.8 LESSONS LEARNED

The lessons-learned report is a key document to share, for consultation on previous experiences, and recommendations on enhancements or modifications to be applied for future projects. The objective is to increase the quality level of management of projects, where previous lessons-learned reports are reviewed to consider how lessons learned from previous projects could be applied to the project.

During the conduct of lessons-learned sessions, participants consider the feedback results to discuss and assess the performance of the project. The lessons may be positive or negative. As lessons learned are documented, the project manager and the team members can develop action plans describing when and how they might be implemented for projects within the performing organization.

8.8.1 Planning for Lessons Learned

As a minimum, lessons learned should be captured at the end of each stage of the project. It is strongly suggested, however, that the lessons-learned process be continuous throughout the project development life cycle. The project manager should plan for the formal capture of lessons learned as part of the regular project progress and review meetings, and where possible, following the rolling-wave planning scheme. Additionally, the project manager must request from team members who have completed their assigned activities, to capture the lessons learned prior to being reallocated to another project, or returning to their original functional departments.

Special attention should be given by the project manager for collecting lessons learned with contractors and providers, especially those who are on the preferred suppliers list. Work performance as well as contractual performance should be reviewed for lessons.

It may also be that during the course of the project implementation, the project manager and team members recognize that the results of certain lessons learned can be incorporated into the ongoing project. These are judged, when exercised, to be effective for improving the production of a deliverable, streamlining a current process, or making recommended modifications to standardized templates.

The lessons-learned report is the final comprehensive document produced during the closeout phase. Suggested content is as illustrated in Figure 8.12.

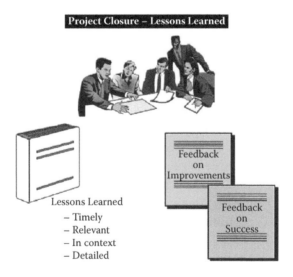

FIGURE 8.12
Project closure—lessons learned

8.8.2 Collating Lessons Learned

As stated in Section 8.8.1, it is the project manager's responsibility to ensure that appropriate meetings and sessions are scheduled to capture lessons learned. The wealth of information that prevails in any project is often lost when either no such sessions exist or the lessons are learned only to implicate those who have failed.

It is of utmost importance that the project manager instills a climate of openness and trust from the first gathering of core team members at the project implementation start-up meeting, and maintains this ambiance throughout the delivery of the project. There is an incoherent contradiction in mapping statements like "getting it right first time" with "you learn from your mistakes." Improvements and enhancements can only be made when there is a willingness to accept that higher quality and effectiveness is a continuous process, during which all are concerned.

The ideal platform for a lessons-learned session is a meeting with project team participants. The group is gathered by the project manager and is constituted by all individuals who have been engaged in any way on the project's performance. When the group size is large, for example over twenty people, the project manager can elect to have subgroup sessions that focus on specific areas of the project's performance. The session leaders can then collate and regroup to consolidate their results with the project manager to prepare the project lessons-learned report.

The suggested table of contents for the lessons learned report should be considered as a starting point for managers to collect and collate lessons learned (see Figure 8.13). In certain cases, the project manager may be requested to focus on certain areas more than others, for example the project management methodology used, or the evolution of team members skills, competences and teamwork dynamics, or even the interrelationship management between the project team, the stakeholders, and key representatives of the performing organization. These focus areas aim at contributing to enhancing and improving project and operational performance.

8.8.3 Checklists for Collecting and Collating Lessons Learned

Depending on the nature of the project, the size and dispersion of the project team, the management requirements and the subsequent use of

Project Lessons Learned Report

	Description
Project Name	
Project Description	
Project Manager	
Project Sponsor	

Table of Contents

1 **Summary**
- 1.1 Project Background
- 1.2 Learning Highlights
- 1.3 Recommendations Summary

2 **Technical Review**
- 2.1 Project Experience
- 2.2 Recommended Process Improvements
- 2.3 Proposed Tool Modifications or Improvements

3 **Administrative Review**
- 3.1 Project Experience
- 3.2 Recommended Process Improvements
- 3.3 Proposed Tool Modifications or Improvements

4 **Contract Management Review**
- 4.1 Project Experience
- 4.2 Recommended Process Improvements
- 4.3 Proposed Tool Modifications or Improvements

5 **Risk Management Review**
- 5.1 Project Experience
- 5.2 Recommended Process Improvements
- 5.3 Proposed Tool Modifications or Improvements

6 **Financial Management Review**
- 6.1 Project Experience
- 6.2 Recommended Process Improvements
- 6.3 Proposed Tool Modifications or Improvements

7 **Customer Relationship Management**
- 7.1 Project Experience
- 7.2 Recommended Process Improvements
- 7.3 Proposed Tool Modifications or Improvements

8 **Team Relationship Management**
- 8.1 Project Experience
- 8.2 Recommended Process Improvements
- 8.3 Proposed Tool Modifications or Improvements

FIGURE 8.13
Lessons learned report TOC

lessons learned, the project manager would be required to develop, or use an existing, lessons-learned questionnaire, which captures the required spectrum of the project and encourages objectivity, especially in areas considered negative (see Figure 8.14).

In essence, for whatever type of questionnaire is prepared all topics need to answer three basic questions: "what went well"; "what went badly"; "what was lacking."

FIGURE 8.14
Collecting lessons learned

The lessons-learned session should include:

- The project manager (who can chair the session or delegate this to a team member)
- Project team member
  - From the internal organization
  - From any contractor or provider
- Stakeholder representation
  - Functional managers
  - Business managers
  - Shared services such as HR, legal, IT, etc.
- Executive management
  - Sponsor
  - Key stakeholders (not included above)
  - Business case owners
- Operations, support, and maintenance groups from the performing organization

8.8.4 Lessons-learned Questionnaire

The focus areas for lessons-learned questions can follow the categories as illustrated for the lessons-learned report:

- Technical
- Administrative
- Contract management
- Risk management
- Financial management
- Customer relationship management
- Team relationship management
- The questionnaire should be drafted to assist in giving objective commentaries on the lessons learned. An approach for rapid assessment can be made by structuring closed questions, for example:

"Did the team use the standard project reporting mechanism?": Y/N.

Other techniques can use a gradient scale from very low to very high, for example:

"How accurate were effort estimations within +/–5% range?"
1=Very low; 2=low; 3=medium; 4=high; 5=very high

Figure 8.15 offers an example of questions.

8.8.4.1 Skills Management

All work performed by team members, during the course of the project, will expose them to new lessons to learn and old lessons to repeat or avoid. Team members' current skills will most certainly be enhanced, and the project manager should ensure that the investment made in improving a team member's skills should not be lost.

The project manager should maintain a record of the skills used and developed on the project and their effective utilization. Such a record will not only facilitate and encourage individual growth, but the skills can be leveraged on future projects. The enhancement of a team member's skills will also be forwarded to the responsible line manager, allowing for subsequent objective performance evaluation.

	Project Lessons Learned Questionnaire

	Description
Project Name	
Project Description	
Project Manager	
Project Sponsor	

	V Low	Low	Med.	High	V High
	1	2	3	4	5
Business Goals Performance					
Business Goals & Objectives were **Smart**					
Business Objectives were met					
Customer was satisfied with the product					
Customer's needs/requirements were met					
Funding and budget were well defined					
Needs & Requirements were documented clearly					
Product concept was appropriate to Business Objectives					
Sponsor Management attention and time to the project					
Stakeholders access to project Plan and Schedule					
Stakeholders commitment and support to project					
Stakeholders input into the project planning process					
Stakeholders satisfaction with reported Information					
. . . .					
Project Delivery Performance					
Change request management was adequate					
Contracted vendors, quality of work					
Documentation was well-controlled					
Procurement management was effective					
Project baselines (Scope, Time, Cost) were well-managed					
Project Objectives were met					
Project plan & schedule documentation was adequate					
Project quality control was adequate					
Project reporting was adequate					
Project schedule encompassed all aspects of the project					
Project tracking accuracy					
Risk management & Issue management were adequate					
Test/Transition plans were adequate & well-documented					
. . . .					
Project Team Performance					
Project manager and team received adequate training					
Project manager was effective					
Project team commitment					
Project team communication					
Project team effectiveness on project goals					
Project team relationship with providers/contractors					
Project team was properly organized and staffed					
Etc. . . .					

FIGURE 8.15
Project lessons learned questionnaire

The project manager should liaise with the corresponding HR department when and if a skills inventory exists within the performing organization. The closure of the project, or the end of a team member's assignment, offers an ideal opportunity for the project manager to capture and discuss with the individual team member the newly developed and/or acquired skills and provide this information to the HR function. An up-to-date skills management inventory can then become invaluable to future project managers for resource assignments and staffing.

If no skills inventory exists within a performing organization, the project manager should, as a key lesson learned, encourage the HR function of the performing organization to implement one.

8.8.5 Exercise: Lessons Learned

Review your current, and at least two previous projects, and prepare a comprehensive lessons learned. Review and assess the results and list the recommendations you can forward to enhance the effectiveness of future projects.

9

Project Leadership Skills

9.1 CHAPTER OVERVIEW

This chapter positions the project manager in the role of the leader. The role and responsibility as a manager and leader of projects set the foundation for the project manager's framework.

The major "soft" skills of leadership and motivation are explained in detail, from which the reader can ascertain how best to create and maintain a dynamic, high-performing project team.

The sections on communication skills and conflict management techniques round off this chapter and will assist the project manager to reach a high level of effectiveness needed to meet the project and corporate goals.

9.2 THE PROJECT MANAGER'S ROLE AS A LEADER

The most usual definition describing the role of a project manager is the person responsible for directing and coordinating the human efforts, material resources, and funding to produce the project deliverables and achieve the goals and objectives. This definition focuses more on the managerial and administrative functions of the project manager, and ignores implicitly the leadership and "people" management facets of the role.

The leadership for project managers focuses on doing "the right things," while acting as a manager focuses on doing "the things right." (See Figure 9.1.)

This results in managing the project and getting things done through the efforts of *people*.

FIGURE 9.1
The project manager's functions

9.2.1 The Role as a Manager

General management skills provide much of the foundation for building project management skills. On any given project, skill in any number of general management areas may be required. Certain general management skills are relevant only on certain projects or in certain application areas. For example, health, safety and environment (HSE) is critical on virtually all construction projects and is of little concern on most software development projects.

Henri Fayol, the father of the school of systematic management, established the five functions of management, which focused on the key relationships between management and its personnel.

1. Planning
   • To establish action plans, according to the organization's resources; type and significance of work and future trends.
2. Organizing
   • To build an organizational structure, and to provide capital, human resources, and raw materials for the operation of the business.

3. Commanding
   - To establish and assign tasks and instructions to employees in the interest of the entire enterprise.
4. Coordinating
   - To unify and harmonize work activities so as to maintain the balance between the activities of the organization.
5. Controlling
   - To control work performance, and to evaluate against plans, policies, and instructions.

The illustration in Figure 9.2 maps Fayol's five functions to Mintzberg's managerial actions.

This sets the scope of the managerial facet of project management and frames the organizational structure for the project manager for doing the "things right" by:

- Administrating
- Avoiding risks

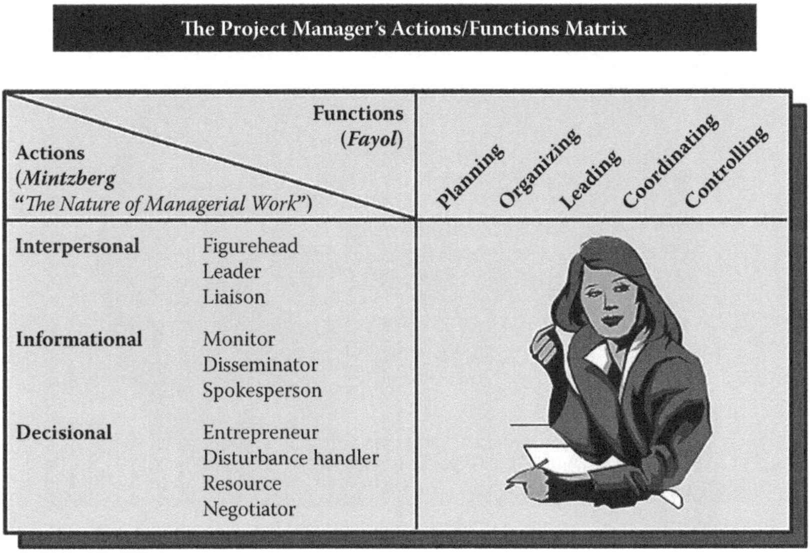

FIGURE 9.2
The project manager's actions-functions matrix

- Conforming
- Directing and controlling
- Instituting procedures
- Measuring financial performance
- Organizing and structuring
- Setting goals and objectives
- Instructing: how and when

9.2.2 The Role as a Leader

Managing projects is primarily concerned with "producing project results expected by stakeholders." Project leadership goes beyond this and focuses on doing "the right things." (See Figure 9.3.)

Project leadership is not confined to the core team members. The project manager needs to demonstrate leadership skills with: stakeholders and

FIGURE 9.3
Managing the project—environment

sponsors, steering committees, upper management, line managers and peers, and external providers.

This is achieved by the project manager's comprehension that leadership:

- Is the process of influencing others in a group.
- Primarily deals with influencing behavior, attitudes, and actions.
- Entails using influence for the project purpose.
- Is bilateral. As a project manager influences project participants, so participants influence the project manager.

Project leadership is an ability to get things done well through others. This requires that the project manager clarify the project goals and objectives to meet; presents a compelling reason to achieve them; provides a realistic timeframe to fulfill them and institutes an attractive and enthusiastic team work environment.

The project manager, unlike a functional manager, performs a variety of actions with little or no line authority.

The project manager's skills in leadership will revolve around the following three major vectors detailed in the following sections.

1. Setting Direction
   - Articulating the vision of the performing organization's future and the project's change strategy for delivering the product/service, to achieve that vision.
2. Inspiring and Motivating
   - Energizing core team members, and other participants, to face and prevail over normal project adversities, such as business constraints; resource limitations; risks and uncertainties; administrative and bureaucratic hurdles; as well as internal and external political pressures.
3. Aligning People
   - Listening to and addressing the needs, concerns, and aspirations of all project participants, and determining the alignment of these to the goals of the organization, the project, the group, and the individuals.

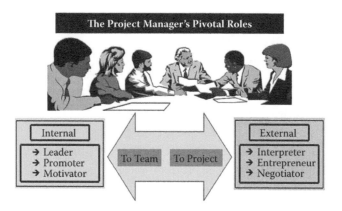

FIGURE 9.4
The project manager's pivotal roles

The project manager has a pivotal role, and his actions branch out inside the organization and extend to external entities involved with the project (see Figure 9.4). The inward focus is to the project core team and requires, above the leadership, a role as a motivator and promoter of the actions and performance of the team members. The outward focus is directed to the major project players such as sponsor and key stakeholders; steering committees and upper management, etc. The project manager is the principal representative of the project to the organization and must demonstrate, not only a keen understanding and interpretation of the key business issues driving the project, but also must act as an entrepreneur to propose alternative choices and avenues in the face of constraints. As a negotiator, the project manager reaches and agrees to decisions with key groups, for matters relating to funding, schedules, resources, contracts, etc.

9.2.3 Effective Project Leadership

The project manager must create an environment that engenders from the core team members participation, energy, involvement, and commitment in the definition, planning, and realization of the project's deliverables and goals (see Figure 9.5). This should also be extended to the sponsor and the key stakeholders.

The project manager's highest priority is to inspire a "Shared Vision" by constantly articulating the essence of the project's existence and how it will contribute to the organization's business benefits. There are many platforms available to do this: at project start-up, regular progress

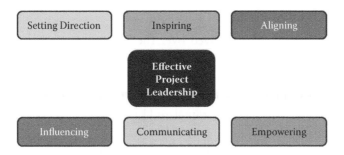

FIGURE 9.5
Effective project leadership

meetings, and other instances when the core team members are present. Flexibility and adaptability are prime requirements for managing projects in a dynamic environment. The project manager must instill with team members the attitude to "challenge the process" so as to reach a balanced way and manner to perform the project without restrictions from archaic and unmanageable processes (see Figure 9.6).

As will be seen in Section 9.2.4.2, Project Manager's Influence Continuum, appropriate delegation will "enable team members to act" as their participation will increase their engagement, commitment, and ownership of their portions of the project. Being part of an interactive and productive team can only "encourage the heart" of project members.

The effectiveness of the project manager as a leader increases with clarity on the project's goals and by setting an example on the attitude to adopt for problem analysis, decision making, risk and change management, quality, and professionalism. The project manager must also demonstrate human traits and characteristics, such as sociability, humor, and gregariousness, as well expressing sentiments, empathy, and compassion.

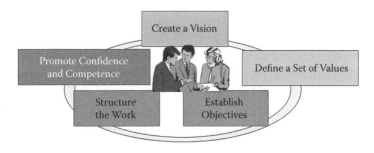

FIGURE 9.6
Leadership with all parties

9.2.4 Leadership Styles

Styles of leadership have evolved over the last two centuries, especially since the beginning of the industrial revolution. A major shift has occurred since World War II, as industries and corporations moved away from primary and secondary sectors to populate the tertiary sector, with the advent of higher-educated personnel and knowledge workers with subject matter expertise.

People management has evolved as a more important focus for managers to achieve higher-quality results in a business environment where societal changes, at least in industrialized countries, have elevated the position of the individual to an important contributing pedestal.

9.2.4.1 McGregor's Theory X and Theory Y

A major contributor to this change is Douglas McGregor who during the 1960s shook the upper-management business community with his Theory *X* and Theory *Y*. McGregor's ideas about managerial behavior, much inspired by Abraham Maslow's need satisfaction model of motivation, had a profound effect on management thinking and practice.

McGregor observations led him to consider that conservative, and what could today be stated as archaic, managerial views led to behaviors and organizational systems that relied on close supervision, rules and regulations, incentives, rewards, threats, and sanctions all designed to control workers. He named this managerial behavior as Theory X:

- A Theory X manager believes and states that, on average, employees really do not want to work. They are self-interested, do not wish to commit to work, would avoid it wherever possible and need direction and control. They do not wish for any responsibility, have little ambition, and prefer a secure and stable life
- Such a manager thus has to apply close supervision and define jobs and systems that structure how a worker must perform
- The relationship between manager and worker is a wage-work bargain, where the technique is to use the "stick" (coercive language, harsh and authoritarian management), and the "carrot" (rewards and promises, which never materialize)

In contrast, and to change the perception and practice of managerial behavior, McGregor promoted a radically different approach that he named Theory Y:

- A Theory Y manager believes that, with the right conditions, employees consider their objectives will complement those of the company, and obtain satisfaction and meaning from their physical work and mental effort.
- Employees relate their efforts to organizational objectives, and exercise self-direction and self-control in the performance of work.
- The Theory Y manager believes individuals show commitment, are capable of handling more complex problems, can and do exercise imagination and creativity, and seek intrinsic rewards associated with their achievements, rather than just extrinsic rewards/punishments.

It may seem that both theories have opposite views as to the participation and ability of the individuals; however, project managers still have the responsibility for organizing the elements of the project plan, engaging team members to cooperate, and aligning the project and the team with organizational goals.

Theory Y is an extension of "job enrichment" (see Section 9.2.5.2, The Herzberg Hygiene-Motivation Theory) and has paved the way to such management concepts as "empowerment" (introduced by Peters in 1982 and 1985). This type of management promotes job development, enabling employees to grow and give more of their innate potential to the corporation.

Whatever leadership styles are used by the project manager, there will always be the challenge of finding the balance between a focus on the work to be achieved and a focus on those who actually perform the work.

9.2.4.2 *Project Manager's Influence Continuum*

The influence continuum, presented by Tannenbaum and Schmidt, is a seven-level scale, along which the decision-making responsibility for the performance of work shifts from the project manager to the team member depending on the knowledge, competence, and skill of the individual. The model depicts the relationship between the level of freedom that a project manager chooses to give to a team and each individual member, and the level of authority used by the manager. As the competence/skill increases,

FIGURE 9.7
Leadership spectrum

so the decision-making autonomy is conveyed to the team member. The different levels of delegated freedom closely relate to the "levels of delegation." (See Figure 9.7.) This is a positive way for both teams and managers to develop. Misuse of the appropriate level by the project manager, either provides too much autonomy to a team member with insufficient competence/skill, or reduces the decision-making ability when a team member possesses the necessary experience.

- The manager decides and announces the decision.
- The manager reviews options in light of aims, issues, priorities, timescale, etc., then decides the action and informs the team of the decision. The manager will probably have considered how the team will react, but the team plays no active part in making the decision. The team may well perceive that the manager has not considered the team's welfare at all. This is seen by the team as a purely task-based decision, which is generally a characteristic of Theory-X management style.
- The manager decides and then "sells" the decision to the group.
- The manager makes the decision as in the previous scenario, and then explains reasons for the decision to the team, particularly the positive benefits that the team will enjoy from the decision. In so doing, the manager is seen by the team as recognizing the team's importance, and having some concern for the team.
- The manager presents the decision with background ideas and invites questions.
- The manager presents the decision along with some of the background that led to the decision. The team is invited to ask questions and discuss with the manager the rationale behind the decision, which enables the team to understand and accept or agree with the decision more easily

than in the first two scenarios. This more participative and involving approach enables the team to appreciate the issues and reasons for the decision, and the implications of all the options. This will have a more motivational approach than the previous two scenarios because of the higher level of team involvement and discussion.

- The manager suggests a provisional decision and invites discussion about it.
- The manager discusses and reviews the provisional decision with the team on the basis that the manager will take on board the views and then finally decide. This enables the team to have some real influence over the shape of the manager's final decision. This also acknowledges that the team has something to contribute to the decision-making process, which is more involving and therefore motivating than the previous level.
- The manager presents the situation or problem, gets suggestions, and then decides.
- The manager presents the situation, and maybe some options, to the team. The team is encouraged and expected to offer ideas and additional options, and discuss implications of each possible course of action. The manager then decides which option to take. This level is one of high and specific involvement for the team, and is appropriate particularly when the team has more detailed knowledge or experience of the issues than the manager. Allowing high-involvement and high-influence for the team, this level provides more motivation and freedom than any previous level.
- The manager explains the situation, defines constraints, and asks the team to decide.
- At this level the manager has effectively delegated responsibility for the decision to the team, albeit within the manager's stated limits. The manager may or may not choose to be a part of the team that decides. While this level appears to gives a huge responsibility to the team, the manager can control the risk and outcomes to an extent, according to the constraints that are stipulated. This level is more motivational than any previous, and requires a mature team for any serious situation or problem. The team must get the credit for all the positive outcomes from the decision, while the manager remains accountable for any resulting problems or disasters.
- The manager allows the team to identify the problem, develop the options, and decide on the action, within the given constraints and limits.

- This is obviously an extreme level of freedom, whereby the team is effectively doing what the manager did in the first scenario. The team is given responsibility for identifying and analyzing the situation or problem; the process for resolving it; developing and assessing options; evaluating implications, and then deciding on and implementing a course of action. The only constraints and parameters for the team are the ones that the manager holds from above.

9.2.4.3 *Situational Leadership*

The Situational Leadership® Theory, developed by Paul Hersey and Ken Blanchard, characterizes leadership styles in terms of the amount of task focus and relationship focus that the project manager provides to team members (see Figure 9.8). The appropriate leadership style to utilize will depend on the person or group under the responsibility of the project manager and the work situation. The main axes of focus are directive or supportive, and are mapped to the individual's competence/skills and commitment.

- **Directing style:** Used with a team member who lacks the specific skills required for the job in hand. The project manager gives direction on how and what to do to perform the task.
- **Coaching:** Used with a team member who is increasing competence/ skills through training, but is still unable to take on full responsibility

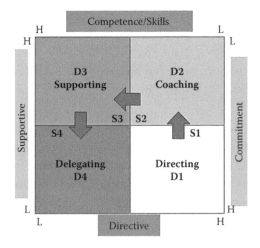

FIGURE 9.8
Situational leadership

for the task being done. The project manager guides the team member to a higher level of performance.

- **Supporting:** Used when the team member has reached the level of competence/skill to perform the task but lacks the confidence to take on full responsibility. The project manager helps the team member through the difficult stage of overcoming the lack of confidence.
- **Delegating:** Used when the team member has attained a high level of competence/skill to perform the task, and is able and willing to take responsibility for the task.

The styles are discussed between the project manager and the team member, and different styles can be used with the same individual, each corresponding to the competence/skill required to perform part or all of the responsibility. For example, the project manager would utilize different situational leadership styles with an experienced project manager with a wide experience on internal project, but lacking exposure on global projects.

In using Situational Leadership styles fittingly, the project manager will:

- Develop the team members' competence/skills and knowledge
- Coach and train team members to reach higher levels of performance
- Support team members who are experiencing difficulties
- Progressively empower team members

The Situational Leadership model, when used in conjunction with the influence continuum, is a powerful and effective way to utilize the appropriate leadership style that is compatible with the competence/skills of the team member and relates to the decision-making level suitable for the performance of the task.

9.2.4.4 Exercise: Situational Leadership

Reflect on your current project and the members of the team and establish for each the most appropriate situational leadership styles to utilize.

9.2.5 Review of Motivational Theories

In psychology, motivation refers to the initiation, direction, intensity, and persistence of behavior—"why we do things." Motivation is a temporal and dynamic state that should not be confused with personality or emotion. Motivation is having the encouragement and willingness to do something. A motivated person can be reaching for a long-term goal, such as becoming a professional program director, or a more short-term goal like learning how to develop a risk management plan. Personality invariably refers to more or less permanent characteristics of an individual's state of being (e.g., shy, extrovert, conscientious), whereas emotion refers to temporal states (e.g., anger, grief, happiness).

In general, motivational theories will refer to the driving forces of an individual, and those forces can be classified into two major groups: obligation to act or willingness to act. Action and drive provoke behavior changes to satisfy a need. Drives and desires activate behavior that is aimed at a goal or an incentive. Basic drives could be sparked by deficiencies, such as thirst, which motivates a person to seek water; whereas more subtle drives might be the desire for impact and recognition, which motivates a person to behave in a manner seeking approval from others.

Drive and behavior change is triggered by unsatisfied needs. Figure 9.9 describes the steps engaged to satisfy a need.

The duration of satisfaction of any need depends on the need itself, thirst for example will be of a short duration and seeking to satisfy it is repetitive in a twenty-four-hour period, whereas impact and recognition may last days, weeks, or even months. Often there is a correlation between the need to satisfy, its physiological or psychological importance, the time and effort to reach it, and the satisfaction duration of the need once achieved.

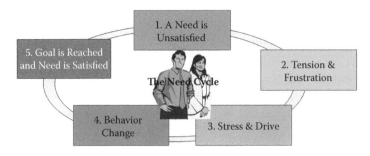

FIGURE 9.9
The need cycle

9.2.5.1 *Maslow Hierarchy of Needs*

No book or article that presents motivational theory can ignore or bypass the work of Abraham Maslow, as it is the most widely discussed theory of motivation. Abraham Harold Maslow was born April 1, 1908, and received his B.A. in 1930, his M.A. in 1931, and his Ph.D. in 1934, all in psychology, from the University of Wisconsin. Maslow noticed, observing and studying monkeys, that some needs take precedence over others. From this observation he created his now famous theory on the hierarchy of needs, where needs provide the driving force of motivating behavior and general orientation.

The theory can be summarized as follows:

- Human beings have needs, wants, and desires, which influence their behavior; only unsatisfied needs can influence behavior, satisfied needs cannot.
- Since needs are many, they are arranged in order of importance, as a pyramid (see Figure 9.10), from the basic physiological to the complex psychological.
- The person advances to the next level of needs once the lower level need is at least minimally satisfied to the person's threshold.
- As the person progresses up the hierarchy, and while the lower obligatory needs are satisfied, the desired psychological needs can be reached.

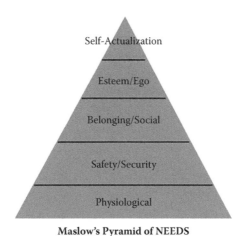

Maslow's Pyramid of NEEDS

FIGURE 9.10
Maslow's hierarchy of needs

The pyramid presents five groups of needs, all of which the individual seeks to satisfy and maintain concurrently, within the available material time. The "lower needs," physiological and safety/security, hold priority and are obligations on the individual to satisfy to an acceptable threshold prior to searching to satisfy the "Upper Needs," the next three levels, which are psychological desires. The challenge for all individuals is how to distribute attention and material time to satisfy and maintain all needs, as all needs are concurrently vying for satisfaction, while lower needs will take precedence when they are no longer satisfied. When an individual is hungry and must eat to subsist, the need to have impact on the social group loses priority to the basic existence need. Each individual will have a threshold of satisfaction per need, which, when met, enables "free time" to search for satisfaction of other needs higher in the pyramid.

- **The physiological needs:** These cover the vital needs for existence: air, drinking and eating, and a minimum intake of protein, minerals, and vitamins. This also covers the needs to be active, to rest, to sleep, and to get rid of wastes. The satisfaction of these needs is compulsory to the individual, as dissatisfaction is terminal.
- **The safety and security needs:** These cover the protection needs for the individual: defense against physical and environmental aggressions; searching for personal safety and stability, securing structure, justice, and order. In general, this means satisfying needs in order to avoid both physical and psychological pain. The satisfaction of these needs is necessary to the individual, as dissatisfaction leads to infirmity.
- **The belonging and social needs:** Human beings constitute social groups, and needs associated with this are important for the psychological balance of the individual. The needs cover friendship, love, children, affectionate relationships in general, and belonging to the community. These needs are more desires than obligations, as on one hand they bring psychological amity while they fend off loneliness and social anxieties.
- **The esteem and ego needs:** As well as seeking belonging and social acceptance, the individual will wish to be recognized as "me—self-esteem." These needs cover two facets. On one side (the view from others) are the respect of others, the need for recognition, appreciation, attention, reputation, status, fame, and glory. On the other side (the individual's own view) are self-respect, confidence, competence, achievement, mastery, independence, and freedom.

Maslow considers these four categories of needs as essentially survival needs, as a satisfied deficient need is no longer motivating. For example, after a very tasty and heavy meal, there is no motivation to eat. This also concerns belonging and esteem as they are needed for the maintenance of health. These four categories form part of "deficit motivation" needs. The fifth category covers a completely different set of needs, which are named "growth motivation" needs.

- **The self-actualization needs:** They involve the continuous desire to fulfill potentials, of becoming the most complete, the fullest, to "be all that you can be—you." This is a level where people have need of purpose, personal growth, and realization of their potentials.

9.2.5.2 The Herzberg Hygiene-Motivation Theory

The work of Frederick Herzberg, as a clinical psychologist, stemmed from his belief that "mental health is the core issue of our times," and he focused on the individual in the workplace.

Herzberg developed the motivation two-factor theory, where people in the workplace are influenced by factors that cause satisfaction or dissatisfaction (see Figure 9.11). Satisfaction refers to the *motivation factors* from the individual's psychological growth, whereas dissatisfaction results from *hygiene factors* from the work environment.

- **Motivation factors** are needed in order to motivate employees into higher performance as they seek "growth from accomplishments." These factors deal with **job content** and lead to job satisfaction. They result from internal generators in employees when they can satisfy needs through personal growth, work itself, responsibility, achievement, advancement, and recognition.

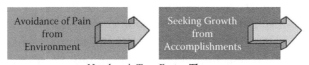

Herzberg's Two-Factor **Theory**
From "The Motivation to work" - Frederick Herzberg

FIGURE 9.11
Herzberg two-factor theory

- **Hygiene factors** are needed to ensure employees do not become dissatisfied as they avoid "pain from the environment." These are factors that deal with **job context**. They do not lead to higher levels of motivation, but without them there is dissatisfaction. These factors include company policies and administration; supervision; interpersonal relations; working conditions; health, safety, and security; and salary.

The key to understanding Herzberg's motivation-hygiene theory is that the factors that involve *job content* (motivation factors) tend to lead to job satisfaction. When these factors are not present on the job, team members do not tend to be dissatisfied—they are simply "not satisfied." Those who are "not satisfied" do not tend to restrict productivity, they just don't get involved in their job or put forth the extra effort to do a good job. Team members who are "satisfied" put forth that extra effort and productivity increases.

Factors that involve *job context* (hygiene factors) tend to lead to job dissatisfaction. When these factors are considered good, or acceptable, team members do not tend to become "satisfied," they simply become "not dissatisfied." Productivity is not restricted—it is just held at an acceptable level. When team members become dissatisfied with any of these factors they tend to restrict their personal productive output.

9.2.5.3 The Application of Motivational Theories

All needs to satisfy compete with each other for the limited time that an individual has to contribute. The lower needs take precedence over the upper needs when they are not satisfied. This leaves the upper needs frustrated as they have no available time to be satisfied. However, self-drive can push back the delay when the lower need kicks in. Working on an exciting report till late at night, against hunger, thirst, sleep, and socialization is an example.

The project manager cannot "impose" motivation on a team member in a sustainable manner. Inspiration lasts as long as the project manager can sustain it; however, sustainable motivation can only come from the team member. The project manager provides the environment and opportunities necessary for people to find motivation in pursuit of goals, development, and achievements that are meaningful to the individual.

Even though there are many derivatives from both Maslow's and Herzberg's theories, they are still the core foundations from which project managers can effectively create an environment in which team members can find motivation. For the project manager, a keen understanding of the hierarchy of needs and the articulation of need satisfaction from the job content, will greatly contribute to effective project work performance.

With reference to Maslow's hierarchy of needs, the project manager can actively influence the satisfaction of team members' needs by providing the following for each of the levels of the pyramid:

- **Self-Actualization**: involvement in planning own work, opportunities for growth, personal development, creative work
- **Esteem/Ego**: freedom to make decisions, status symbols, recognition and awards, challenging work, opportunity for advancement
- **Social/Belonging:** opportunities to interact/network, team-based work; friendly co-workers
- **Safety/Security**: job security, sound policies and practices, proper supervision, safe working conditions
- **Physiological:** adequate compensation; rest periods

With reference to Herzberg two-factor theory, the project manager can influence greatly the job content, while ensuring that the job context, often outside the project manager's field of control, does not outweigh with negative hygiene factors the positive motivation factors.

- Job content—motivational factors
  - Growth—provide learning of new skills, with greater possibility of advancement within the current occupational specialty, as well as personal growth
  - The work/task—provide interesting, varied, creative, and challenging assignments
  - Responsibility—provide appropriate responsibility and authority in relation to the assignment with respect to the agreed upon leadership style(s)
  - Achievement—provide tangible and objectively measurable means to evaluate and render visible the team member's results
  - Advancement—provide for promotion and mobility in the company
  - Recognition—provide for recognition for an assignment well done or the achievement of a team member's personal accomplishment

- Job context—hygiene factors
  - Company policies and administration—avoid and/or protect against inadequacy of company organization and management by adapting procedures and rules to the content of the work assignment
  - Supervision—ensure effectiveness and competency in the direct supervision of team members
  - Interpersonal relations—promote open and bilateral assignment related interactions and social interactions within the project environment
  - Status—ensure that team member's indication of status is recognized and displayed, such as: job/position title, private work space, company car, etc.
  - Working conditions—provide clean, comfortable, safe, and healthy physical work environment for the assignment
  - Job security—protect against job and/or assignment instability
- Compensation—provide for fair salaries and other forms of remuneration

9.2.5.4 Exercise: Motivation

Reflect on your current project and determine which areas you can improve to create an environment in which team members can best satisfy their motivational needs and reach higher levels of performance. Also list those environmental and company factors that you believe are causes for dissatisfaction.

9.3 THE PROJECT MANAGER AND THE ORGANIZATION

The project manager will continually need to influence the organization to "get things done." This will be internally with the core team members or with the sponsor and key stakeholders and will extend to all external entities, such as the customer/client and providers and contractors. This will require knowledge of the formal and informal structures of the organizations involved, principally the performing organization, and an understanding of the mechanics of power and the prevailing politics (see Figure 9.12).

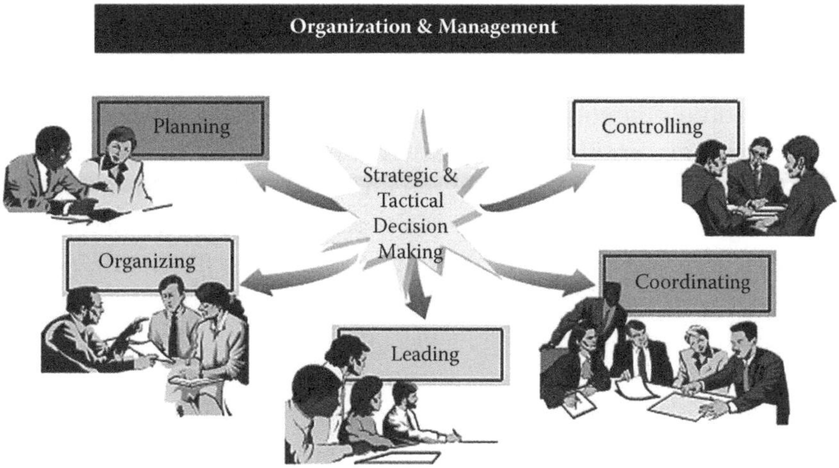

FIGURE 9.12
Organization and management

Power in this context refers to the ability to influence behavior and to overcome resistance (for more details, see Section 9.5). Politics is engaging collective action from a group of people who may have quite different interests. If incorrectly applied, politics will degenerate into power struggles and organizational conflicts leading to inefficiencies and counter-productivity.

The areas the project manager needs to continually focus on are:

- The project core team
  - Fulfilling project goals
  - Defining roles and responsibilities
  - Engaging core team members
- Sponsor and stakeholders
  - Securing commitment
  - Positioning stakeholders in the project environment
  - Understanding stakeholders' agendas
  - Networking with stakeholders
- Providers, contractors, and suppliers
  - Relationship management
  - Defining participation of external providers
- Establishing bilateral communications

9.3.1 Managing the Project Team

Once the project is authorized and a charter is prepared, the project manager constitutes the core project team, where members are assigned from the performing organization and external providers, in the case of outsourcing and contracting (see Figure 9.13).

The organization and structure of the project and the subsequent distribution of responsibilities to core team members is related to the nature and scope of the project. This has been addressed in previous chapters, and covers defining roles and responsibilities for:

- Project planning and control
  - Scope performance
  - Schedule performance
  - Financial performance
  - Tracking
  - Progress and status reporting
- Risk management
- Change management
- Communications management
- Procurement management
- Quality management
- Meetings schedule

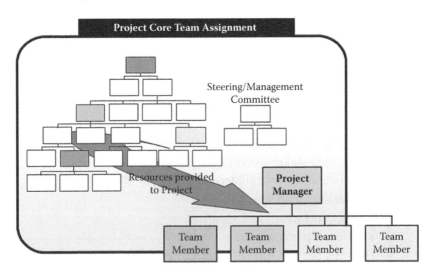

FIGURE 9.13
Project core team assignment

- Interfaces to other projects
- Partner management
- Contracts management
- Project administration

9.3.2 Team Dynamics

Group dynamics are the characteristics of groups and/or teams that are studied in order to analyze the effectiveness or dysfunction of the group or team. At the outset, a group of individuals is not as yet a Team (upper-case T), and remains as a collection of persons who, for the ease of use, are called a team (lowercase t).

9.3.2.1 Moving from a Group to a Team

At the constitution of the project team, the project manager is faced with transitioning from a group of individuals to a dynamic and turned-on effective team. Social groups get things done successfully through collaboration for a common cause, and this has been demonstrated throughout history. The catalyst is the person responsible for the destiny, structure, and advancement of the group. The project manager must fulfill that role fully.

General definitions of a group and of a team:

- A **group** is a number of persons, usually reporting to a common superior and having some face-to-face interaction, who have some degree of interdependence in carrying out tasks for the purpose of achieving project and organizational goals.
- A **team** is an active group of people committed to common objectives who work well together and enjoy doing so, have complementary skills, work interdependently to achieve specific performance goals, and hold themselves accountable for the results. Multilateral exchanges and the communication fluidity in the interrelationship of members of a social group is a prime factor for team effectiveness. This is hampered when the size of the group surpasses nine persons. A group of one hundred employees cannot be considered a team, nor can a sports team with a large squad size. However, in the vernacular, they are called teams.

A project team is a social group, and Bruce W. Tuckman, an educational psychologist, first described (in 1965) that groups will go through four distinct stages of development before they achieve maximum effectiveness. The model was expanded to five stages in 1977.

Too often project managers assume that a newly constituted project team will "hit the deck and start running" from day one, with no questions to be asked, and head for the assigned deliverables. This has proved disastrously wrong on many occasions.

9.3.2.2 The Stages of Social Group Development

On observing behavior of small groups in a variety of environments, Tuckman recognized that the group must experience all four stages in a set sequence, with of course the possibility to step back to a previous stage when conditions dictate this.

As Figure 9.14 shows, the stages are linear and, when used effectively, can move the group to be a Team.

Work on the project is, of course, being performed. However, until stage 4 is reached, the group can still not be considered as a performing team. The project manager must allow time and space for each of the stages and not directly go from stage 1 to stage 4. Each stage is further explained in

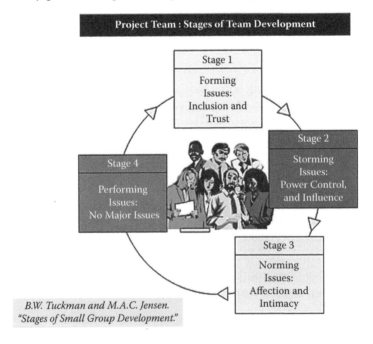

FIGURE 9.14

Project team—stages of team development

the following sections, along with the actions that the project manager must engage or perform.

9.3.2.2.1 Stage 1—Forming

Forming is the initial stage of the group formation, and intervenes usually at the start of the project when the core team members have been assigned. The first meeting of all the team members is usually at the project start-up. The forming stage is also used when regrouping stakeholders for a first meeting, or even external providers. The stage initiates the constitution of the social group.

Individual behavior is driven by a desire to be accepted by the others, and to avoid controversy or conflict. During the first encounters serious issues and feelings are avoided, as people wish to focus on the team organization, goals and objectives to meet, the scope of the work to be performed and the constraints, etc. However, individuals are gathering information and impressions about each other.

With reference to the leadership and motivational aspects as described in the previous sections, it can easily be understood that all participants to this first meeting will demonstrate and feel a mixture of excitement (of something new) and anxiety (because it is new). There are many unknowns, and there are new faces around; some may also feel possible boredom or cynicism for those who have unfortunately been through projects that never reached a stage of a performing team. All are in an expectancy mode to hear and listen to the project manager and receive information on the project, their roles, and other issues that they wish to be comfortable with. Most if not all the attendees would be hovering between the safety/security and social/belonging levels of the hierarchy of needs. At issue are factors of inclusion and trust.

This stage is usually a project or phase start-up (kick-off as it is also called). The project manager's focus in this meeting is to very quickly establish a rapport and an encouraging atmosphere, and the manager must spend the opening part of the session with introductions and ice-breakers. The members of the audience are still individuals and they reason as "I." The project manager has still not engaged the group, nor do the attendees feel part of the project as yet, be it psychologically or physically. Therefore, the project manager should avoid at this stage using the "We," as there are no "We's" as yet. Also the project manager should avoid a personal use of "I," as this shifts the group's attention away from the project and to the project manager. The use of neutral third person singular,

"It," elevates the group's attention to focus on the project. So rather than say "My objectives are to deliver this project by the end of the year and to this budget, and I would need your total support throughout the project to meet the company's goals," the phrase should be formulated as "The project goals are to deliver by the end of this year to this budget, and the participation and collaboration of all team members will be a key success criteria to meet the company's goals."

At this stage, the project manager has a somewhat captive audience, and must present the project and its information in a clear, comprehensive, coherent, and exciting manner. The project manager has a platform to evoke the guiding philosophy pertaining to such things as performing a project in an environment of uncertainty, the quintessence of team dynamics, the challenge of change, and must be eloquent at least on the following:

- Origin of project and organizational goals to meet
- Project organization, roles, and responsibilities
- Clarity of the links between accomplishments and goals
- Processes and standards for the project
- Performance measurement and evaluation mechanisms
- Communication and reporting techniques
- Decision making and escalation procedure

9.3.2.2.2 Stage 2—Storming

This is a key stage to integrate group members, as it allows each and one to clarify and seek answers to areas of doubt.

Individuals in the group can only remain restrained and silent for so long, as issues they consider to be important begin to appear. These may relate to any aspect of the project, the work of the group itself, to roles and responsibilities within the group, or any other issue the individual wishes to clarify and or rebuff. Depending on the culture of the organization and individuals, and most importantly the attitude of the project manager, the issues will be more or less suppressed, but will remain unsettled.

Contrary to stage 1, during which the project manager held the majority of the attention and focus on the project, this stage is now for the group members to express and voice their issues. The project manager must allow this to happen, and this is often done either at the end of the meeting by inviting questions and commentaries, or by planning, for projects with substantial information to digest, a follow-up meeting specifically designed to clarify doubts and answer questions. In certain cases, individuals may wish for a one-on-one with the project manager to discuss personal or professional issues in private.

Storming is of great importance to the subsequent effectiveness of the team, and the project manager must be patient to allow all team members to express themselves and to challenge key project parameters and constraints. Failure to do this will nurture dissatisfaction, which will have impact on both the individual's performance and the group's dynamics. The project manager must therefore be disposed to:

- Listen to issues
- Confront difficulties
- Manage conflicts
- Resolve organizational roles and responsibilities frictions
- Provide support: skill development, resources, access to information, etc.
- Help team to make decisions

9.3.2.2.3 Stage 3—Norming

This stage is indicative of the group's capability to establish its working model and to progressively evolve to a Team. The project manager should guide the group out of stage 2 by reaching agreement on the issues that have been raised. There are many indicators that the group has entered this stage, namely that members have acquiesced on the distribution of work, tasks or responsibilities are clear and agreed upon, methods of work and reporting are suitable, communication methods are acceptable, etc.

The project manager can measure rapidly the growing level of involvement and commitment by the project members by listening for a key word spoken during group discussion: "We." This short word, or variations to it, attests that members now understand each other better, listen to each other, appreciate each other's skills and experience, appreciate and support each other, and are prepared to work together. This is a major step to establishing a cohesive, effective group. The project manager reinforces the above, so as to exit this stage and enter stage 4:

9.3.2.2.4 Stage 4—Performing

This stage is characterized by a state of interdependence and flexibility, where project members work as a Team. Members know each other well enough to be able to trust each other and work together and independently. Group identity, loyalty, and morale are all high, and the project team is task-orientated and people-orientated. Team energy is directed towards the project and its goals.

The project manager does not now "rest on the laurels," but maintains and sustains the team dynamics, by monitoring the project progress and the team's evolution. Corrective actions are easier to introduce at the stage, as the team members have a high sense of commitment and engagement.

9.3.2.2.5 Stage 5—Adjourning (Closeout Stage)

This stage was introduced by Tuckman some ten years after first describing the four stages.

The stage concerns the team member's work completion and disengagement from the project and the group members. The project manager must recognize the sense of loss felt by the team member, and ensure that the individual's departure is celebrated in an appropriate way, for example a farewell lunch or dinner. The project manager must also complete whatever has been agreed for the performance evaluation of the team member and forward this to the respective manager or supervisor.

When the project ends, the team of remaining members is demobilized. The project manager follows the same steps as above. However, the end of project evaluation, recognition and celebration should be made such that members experience an important event.

9.3.2.2.6 Final Comments

Tuckman's original work described the way he had observed groups evolve, whether they were conscious of it or not. For the project manager, the real value is in recognizing where a group is in the process, and guiding it to move to the Perform stage of an effective Team. Another area of focus for the project manager is the dynamics of change in the membership of the team. Project team groups are often forming and changing according to the assignment and completion of work. As this occurs, the group shifts to a different Tuckman Stage. A group might be in the Norming or Performing stage—the arrival and or departure of a member may well oblige the group to return to the Storming stage. The project manager must anticipate and recognize this change of situation, and help the group get back to Performing as quickly as possible.

9.3.2.3 The Performing Team

The primary aim of the project manager is to establish the environment where project team members can effect their assignments in a professional and social atmosphere that provides them the factors to satisfy their motivational needs. As seen above, the project manager needs to combine the knowledge and practice of leadership, motivation, and team dynamics to raise the competence and confidence of team members to become, maintain, and sustain a performing team.

Team members will perform to their highest motivated levels when they feel comfortable that they will succeed in completing the assigned work. Being part of a performing team, where the social group is in harmony and communication is open and honest, raises the team member's appreciation of group membership and provides opportunities to have a personal impact.

In short, the team member satisfies the upper motivational needs as described by the Maslow pyramid of the hierarchy of needs.

The project manager must be aware of the factors that can disturb a performing team and render it as a dysfunctional team. These factors are:

- Absence of mutual trust
- Fear of open/direct conflict
- Lack of commitment
- Avoidance of accountability
- Inattention to results

9.3.2.4 Giving Meaning to the Task

The project manager must give a meaning to the task by ensuring that the team member is able and capable of performing to the required level of quality. Figure 9.15 shows the iterative three steps that the project manager engages with each team member.

The three steps of planning, execution, and control are aligned to the agreed situational leadership established between the project manager and the team member.

9.3.2.5 Effective Team Characteristics

There are numerous characteristics that demonstrate that the Team is effective and "turned-on." Simply said, "Listen to the stories they tell and

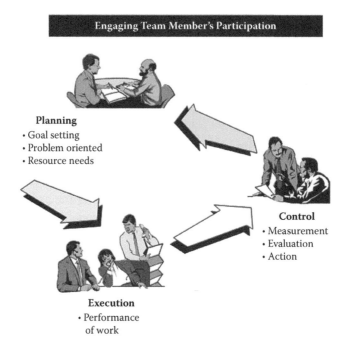

FIGURE 9.15
Engaging team member's participation

look at the smiles of their faces," though not very scientific, summarizes the list below (in alphabetic order):

- Civilized disagreement: There is disagreement, but the team members are comfortable with this and show no signs of avoiding, smoothing over, or suppressing conflict.
- Clear purpose: The vision, mission, goal, or task of the team has been defined and is now accepted by everyone.
- Clear roles and responsibilities and work assignments: There are clear expectations about the roles played by each team member. When action is taken, clear assignments are made, accepted, and carried out. Work is fairly distributed among team members.
- Consensus decisions: For important decisions, there may not be a unanimous agreement. The team reaches consensus through open discussion of all members' ideas, avoidance of formal voting, or easy compromises.

- External relationships: The team members engage in developing key relationships, mobilizing resources, and building credibility with important project players in other parts of the organization.
- Informality: The climate tends to be informal, comfortable, and relaxed. There are no obvious tensions or signs of boredom.
- Listening: The members use effective listening techniques such as questioning, paraphrasing, and summarizing to get out ideas.
- Open communication: Team members feel free to express their feelings on the tasks as well as on the way the group functions. Communication continues outside of meetings.
- Participation: There is much discussion and everyone is encouraged to participate.
- Self-assessment: Periodically, the team stops to examine how well it is functioning and what may be interfering with its effectiveness.
- Shared leadership: While the team has a formal leader—the project manager—leadership functions shift from time to time depending upon the circumstances, the needs of the group, and the skills of the team members. The project manager displays the appropriate behavior and guides the establishment of positive norms.
- Style diversity: The team has a broad spectrum of team "player types."

9.3.2.6 Exercise: Team Dynamics

Reflect on your current project and determine in what development stage the project team is presently and identify what actions you can and need to take to move the group to a performing team.

9.4 THE PROJECT MANAGER'S COMMUNICATION

The project manager must go beyond establishing a communications plan, which is only the mechanism and process for the collection and dissemination of information. A deep knowledge of how and what to communicate on a personal basis is primordial for the project manager (see Figure 9.16).

FIGURE 9.16
Communicate

This section describes the foundations and techniques of person-to-person communication that are essential for the management of team members, sponsors, stakeholders, external providers, and all other project players. The first aspect that must be well understood is that a large percentage of communication is not verbal (see Figure 9.17).

Language is the representation of experiences, where messages are generalized, filtered, and distorted. Perception is the way information is extracted from the surrounding world. Sensory input is correlated to personal experience.

- Communication problems to overcome
  - Personal representative models of the world
  - Filtering what is seen and heard to conform to personal needs
- Tendency to see/hear what is thought to be seen/heard

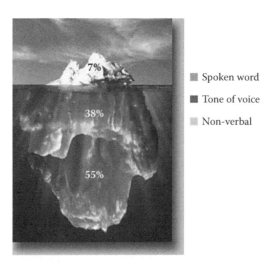

FIGURE 9.17
Nonverbal communication

9.4.1 Communication Basics

The classical communication model is the Shannon-Weaver Model (1947), which proposes that all communication must include six elements: a source, an encoder, a message, a channel, a decoder, and a receiver. The model was initially developed to address technological communication. The human communication model in Figure 9.18 embraces a larger spectrum of factors.

The emphasis here is very much on the transmission and reception of an information message. The message transmitted consists of a core, which may be incomplete with the omission of certain facts, while additional information is included. The message will be composed with a rational and emotional part, depending on the state of the sender, and is transmitted verbally and with the other senses. The message transits via a communication medium, which may distort its contents—this is the noise. For example, saying "Good morning," while shaking a hand firmly and smiling, uses multiple message components, to which tone of voice and inclination can be added. Distortion can happen when there is discord between the message components themselves, for example saying good morning but with a stone face, or more often an ambient disturbance, for example the passing of a noisy car that may drown the volume of the message.

However, the most important factors reside in the states of both sender and receiver. The formulation of the message has its roots in the inner perceptions and experiences, the filtering due to culture, and the emotional level of the dialoguers, as verbal and body language paint emotions.

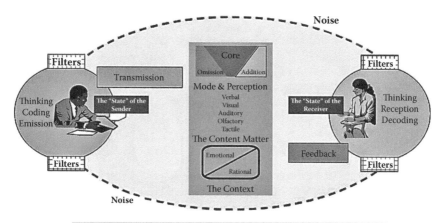

*A Synthetic Model representation, expanded from the Shannon-Weaver Model

FIGURE 9.18
The message interchange model

The transmission of visual, tactile, and auditory information is rooted in the group culture. For example staring, handshaking, or touching may be prescribed, and level or pitch of voice may vary.

According to Mehrabian (Silent Messages, Wodsworth, 1971), transmission and reception consider more than words. It is most important to understand that a large percentage of communication is through: body movements and posture, gestures, eye contact, touch, space, voice. Attitudes (behaviorism) are also communicated by body language signals; however, these must be taken in context. The most common are:

- Confidence: lowered shoulders, hands behind your back
- Evaluation: frowning, hands on hips
- Defensive: crossing arms or legs; raising finger
- Frustration: biting of the lips, clutching the fist
- Insecurity: biting nails, playing with your hair

It should be noted that the majority of body language signals are coded/decoded faster than the cognitive formulation time, by an average of 100 milliseconds. Other signals may have their roots in habits and culture. Furthermore, there exist six universal facial expressions that give clear signals (beware that these may vary depending on culture). These are seen in Figure 9.19.

FIGURE 9.19
The six expressions

EXPRESSION	MOTION CUES
Happiness	Raising and lowering of mouth corners
Sadness	Lowering of mouth corners, raising inner portion of brows
Surprise	Arching eyebrows, while eyes open wide to expose more white, and the jaw drops slightly
Fear	Raising eyebrows, with open eyes, and mouth slightly open
Disgust	Raising upper lip, wrinkling nose bridge, raising cheeks
Anger	Lowering eyebrows, with lips firmly pressed, and eyes bulging

The Shannon-Weaver Model examines the different components that constitute a communication exchange. These are explained in the following sections.

9.4.1.1 The Source—Sender

All human communication has some source, some person or group of persons with a given purpose, a reason for engaging in communication (see Figure 9.20).

9.4.1.2 The Encoder

In person-to-person communication, the encoding process is performed by the motor skills of the sender: vocal mechanisms (lip and tongue movements, the vocal cords, the lungs, face muscles, etc.), body language, such as facial muscles, and so on.

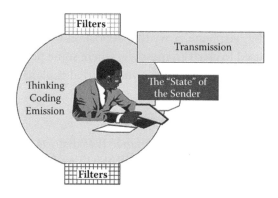

FIGURE 9.20
The sender–message construction

FIGURE 9.21
Message constituents

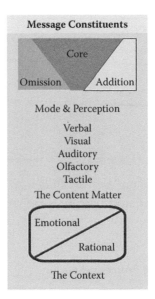

9.4.1.3 The Message

The Shannon-Weaver Model separates the message from other components of the process of communication. However, the message must be examined within the context of all the other interlinked elements. Communication is circular and involves sending and receiving messages, each with an initial feedback. The feedback verifies to what extent the message received corresponds to the message transmitted (see Figure 9.21).

9.4.1.4 The Channel—the Medium

The choice of the appropriate medium is vitally important in communication: this could be a gesture, a vocal verbal exposition of an argument face-to-face or by telephone, a written report, etc. At times, more than one medium is used to clarify a message, where some people may draw or sketch a representation to elucidate a vocal verbal explanation.

9.4.1.5 Physical Noise

Physical noise is any ambient disturbance that distorts the message. This is very often related to the medium used. The impact of the physical noise may create a mismatch between the intended sent message and the interpreted received message. This leads to what is normally referred to as semantic noise (see Section 9.4.1.6).

9.4.1.6 Semantic Noise

Semantic noise is different from physical noise and is not just an ambient disturbance. The coding/decoding of the message may be distorted with relation to people's knowledge level, their communication skills, their experience, their prejudices, cultural background, beliefs, etc. Examples of semantic noise would include:

- Incompatible coding/decoding convention: Both the sender and receiver use words or gestures that are not in their repertoire. For example using esoteric language or a local/national signal, like the shaking of the head to show agreement.
- Attitude towards the other: Preconceptions on the status, rank, origin of the individual, which leads to formulating assumptions on the appropriate coding/decoding to use. For example, on the following occasions: communicating as a project manager to a team member, exchanging with upper management, representing the company in the presence of an external provider, dealing with a non-national on a global project, etc.
- Message content: Acceptance and respect of the received message, while rejecting the contents as being unconvincing or in disaccord with beliefs. For example, communicating on political, religious, or philosophical matters.
- Message focus: The message content emphasizes peripheral issues and does not address the core topic.
- Distraction: The receiver's attention is directed elsewhere, be it mental or physical, for example a fleeting thought, or a moving object. There is no physical noise that prevents the message from reaching the receiver. However, the distraction blocks part or all of the message decoding.

9.4.1.7 The Decoder

This is the reversal process of the receiver to decode/retranslate the sender's encoded translation of the purpose of the message. Communication breakdown can occur on the assumption that receivers decode messages in the same way as the sender.

9.4.1.8 The Destination—Receiver

For communication to occur, some person or group of persons must be present to receive and decode the message, and utilize a compatible system

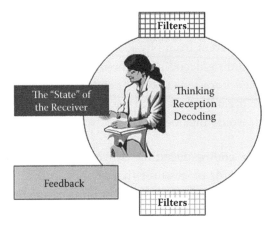

FIGURE 9.22
The receiver–message decoding

of communication. For written messages, the receiver must not only be able to read, but must also be conversant in the language used (see Figure 9.22).

9.4.1.9 Feedback

The feedback is formulated by the receiver and sent to the original sender as a sign of either comprehension or incomprehension of the message and its contents. This is an important step in the process if the communication is to proceed to address the original purpose of the sender. When the feedback indicates comprehension, then both the sender and receiver can move to the next message, else the sender would need to reformulate the message. For example, a mumbled word spoken on the telephone may be incomprehensible to the receiver who may feedback to the sender to repeat the word.

9.4.2 Communicating Techniques

The project manager, being the fulcrum of the project's communication network, must constitute a large palette of techniques and corresponding skills to ensure that the appropriate and targeted messages flow between all project players. These messages must convey clear and comprehensive contents in a context that should not allow distortions.

The hierarchical and organizational position of the project manager sets a communication standard for the project. The way the project manager

FIGURE 9.23
The project manager's communication with stakeholders

communicates is continuously evaluated by team members, sponsor and stakeholders (see Figure 9.23), as well as representatives of the performing organization and external providers.

The project manager's personality traits are perceived and quickly assessed by observers of the project, rightly or wrongly. To avoid as much as possible incorrect interpretations, the project must bond and connect with people and seek commonalities by communicating clearly and frequently. The project manager must show integrity and be honest, sincere, and ethical.

The project manager builds rapport with all project players and maintains and sustains this with open communication, a positive attitude of interest in the other person, and exploration of their interests. The overarching goal is to seek and establish commitment, involvement, engagement, understanding, and awareness. The project manager should make a conscious effort to assess the communication techniques used and constantly evaluate their effectiveness and must strive to:

- Be aware of the style of communication used
- Be open to new information
- Put effort into listening
- Make an effort not to be defensive
- Let the person know they are understood
- Listen for what you can be understood

- Develop a positive attitude
- Be aware of feelings, emotions
- Pay attention to personal, cultural, and/or other differences
- Avoid being judgmental and critical of others
- Treat all persons with respect

The project's communication plan establishes the formal network and vehicle for the circulation of information from and to project players. Depending on the nature of the project, the manager should analyze the plan and modify and/or adapt it to correspond to the communication goals sought.

Different formats are needed for different communication objectives, where some are more effective and essential than others. Too often, electronic messages are used for the wrong reason. For example, the resolution of a personal conflict is not to be addressed by e-mail and should be done face-to-face in a place that provides privacy for open and frank discussions; a long and complicated design exposé requiring critique should not be presented in a meeting but should first be distributed to allow time for the reader to understand the contents before forwarding comments and remarks, etc.

Communication formats can be grouped under face-to-face, hard-copy, and electronic. Each will have different message contents and contexts, and when ill-prepared will potentially introduce noise and distortion. Here are some of the communication vehicles by format:

- Face-to-face
  - Project team meetings
  - Planning workshops
  - Senior management meetings and visits
  - Cross-functional meetings
  - Interactive staff meetings
  - Personal issue resolutions
  - Away days and conferences
- Hard-copy
  - Newsletters
  - Business plans
  - Letters and memos
  - Circulars

- Electronic
  - Telephone
  - Video and audio conferencing
  - Web-enabled exchanges
- Computer networks and emails

The project manager, like all project players, is both sender and receiver of messages. What must be considered is that the collection, synthesis, and dissemination of most if not all of the project information is performed by the project manager.

The sources of information and their destinations vary depending on the individuals involved and the frequency of communication exchange (see Figure 9.24). For example a project manager responsible for a team of four people, all sitting and working in the same office perimeter will communicate very often, orally, and spontaneously. This is different for

FIGURE 9.24
Communication network

a large project launched by a consortium and where written reporting is very formal and at a quarterly frequency.

9.4.2.1 The Project Team

The project manager communicates greatly and most frequently with the project team, and has regular opportunities to articulate and fulfill the different roles of leader and motivator by:

- Stating group objectives
- Explaining goals, processes, and standards
- Keeping team members aware
- Discussing differences of view
- Offering constructive criticism
- Supervising performance and praising
- Giving instructions and maintaining discipline
- Resolving conflicts and grievances

9.4.2.2 Sponsors and Stakeholders

The communication plan and the nature of the project dictate both the contents and frequency of information exchange with the sponsors and stakeholders. In most cases, the frequency would be monthly, if not quarterly or even half-yearly. The project manager must assess the impact this has on the relationship management between the parties, and how messages must be finely constructed to be targeted and correctly interpreted.

Sponsors and stakeholders have operational or strategic objectives as well as professional goals and motivations. Furthermore, different stakeholders may have different and conflicting interests and expectations and their level of participation differs. Executives and line managers have positive/negative financial or emotional interest in the outcome of their performance, and have key motivational drivers. Professional and personal interests will influence the project and will often go beyond the project's goals. These personal goals and political motivations impact on decision making. The project manager must thus be aware and conscious of potential message distortions.

9.4.2.3 Performing Organization and Peers

The project manager has also to construct an effective relationship network with managers and representatives of the performing organization.

Line managers have operational responsibilities and may not be focused on the project throughout the development life cycle, until such time that they realize that the product/service is ready for acceptance and handover. Organizational readiness is crucial to fulfill the company's operational goals, and the project manager must establish a convenient frequency of meetings with the respective managers. The project manager must anticipate that line managers and their staff will demonstrate different levels of resistance, depending on their acceptance or rejection of the proposed change the project will create. Care must therefore be taken by the project manager in the communication techniques to use to guide the performing organization representatives to an unreserved commitment.

Peers will comprise mostly of other project managers. They all are contending for limited funding and resources, and the progress of their own projects has a direct impact to the project manager's own project. Frequent meetings at scheduled milestones and other interface points should be established. The communication must be open, frank, and collaborative.

9.4.2.4 External Organizations—Providers

The project manager is the company's representative to external organizations, such as providers, contractors, and suppliers. The project manager will be regarded as a client, and the behavior of external managers will often reflect some form of subordination in the client-supplier framework. This type of behavior may introduce distortion and noise in the communication relationship between the parties. The project manager must be conscious of this and demonstrate equity in the relationship and avoid any superior stance that will compound the message distortions.

9.4.4 Styles of Communication

As has been explained above, the mode and structure of communication and the messages that transit the information paths are fraught with possibilities of misunderstanding and incomprehension and are subject to both physical and/or semantic noise. The project manager must gauge the styles and impact of the communication techniques used (see Figure 9.25).

In all communication exchanges, human signals are interpreted through perception with anticipatory expectations, by assessing nonverbal (body

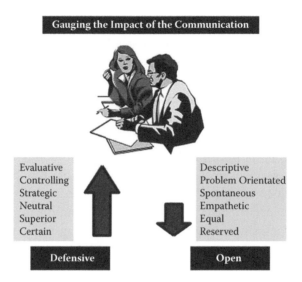

FIGURE 9.25
Gauging the impact of the communication

language) and verbal (grammar, tone, and volume) and how they match the oral/written language that accompanies the message.

As illustrated above, there are a variety of circumstances, needed for the fulfillment of the job, during which the project manager may implicitly or explicitly provoke an undesired reaction and feedback. In instances when the project participant feels threatened, for example a project progress meeting, defensive communication is triggered and barriers are raised. The project manager must anticipate this and instill a climate of confidence and trust whenever the communication exchange excites this kind response.

9.4.5 Listening

Of all the communications skills, listening is arguably the one that makes the biggest difference. Listening does not come naturally to most people, as they tend to be more interested in announcing own views and experiences than really listening and understanding others. "Seek first to understand, and then to be understood," serves as a constant reminder for the need to listen to the other person before expecting to be listened.

Listening is rarely confined merely to words, and in its fullest sense, ultimately includes many nonverbal and nonaudible factors, such as body

Ear   You   **FIGURE 9.26**
Eyes   Listening—all senses
Undivided
Attention

Heart

language, facial expressions, reactions of others, cultural elements, and the reactions of the speaker and the listeners to each other.

Active listening and empathy are interrelated and imply listening with eyes, ears, and body (see Figure 9.26).

There are two general categories of listening. Positive listening and its variations are honest behaviors demonstrated by the listener. Insincere listening and its variations may seem to show interest in the speaker; however, when unveiled they show contempt.

- Positive listening and variations.
  - Active listening—understanding feelings and gathering facts.
- Listening to words, intonation, and observing body language and facial expressions, and giving feedback without emotional involvement, or empathy. There is no transmitted sympathy or identification with the other person's feelings and emotional needs. This listening gathers facts and to a limited extent feelings, too, but importantly the listener does not incorporate the feelings into reactions.
  - Empathic listening—understanding and correlating facts and feelings.
    - Listening with full attention to the sounds, and all other relevant signals, including tone of voice; other verbal aspects (pace, volume, flow, style), body language.
    - Showing appreciation of how the other person is feeling.
    - Recognizing cultural, ethnic, or other aspects of communications and signals used by the person.
    - Giving feedback and verifying understanding with the speaker.
    - Honestly expressing disagreement along with genuine understanding.
  - Facilitative listening—understanding fully, and helping to meet the other person's needs.
    - Extension of empathic listening as help to the other person.
    - Interpreting the self-awareness of the speaker.
    - Restraint of making decisions on the other person's behalf.

  - Showing interest in helping the other person see and under-
    stand their options and choices.
  - Devoid of any selfish personal motive.
  - Facilitative listening is an attitude of mind.
- Insincere listening and variations.
  - Passive listening—ignoring
    - No concentration and no registration.
  - Responsive listening—insincere active listening.
    - Demonstrating the external signs of active listening, while
      internally it is passive listening.
  - Selective listening—biased and projective, intentional disregard
    of the other person's views.
    - Listening to content only, and failing to capture all nonverbal
      sounds and signals.
    - Gathering reliable facts, but failing to gather and respond to
      emotions and feelings.
  - Interpretative listening—unconsciously overlaying own views
    with invalidated assumptions.
    - Showing interest to received information.
    - Interpreting content to suit own purpose.
  - Manipulative listening—fact gathering and analysis for own
    purpose.
- Using all the above insincere listening techniques to arrive at per-
  sonal ends.

9.4.6 Guidelines for Communication

The project manager must take up the responsibility for the success of the
communication and be aware of the exchange of verbal and nonverbal
signals, and to rapidly establish an honest and sincere climate and reduce
emotions from the message when these are perceived (see Figure 9.27). The
project manager has certain basic rules to follow, such as concentration on
the message, no distraction, identifying the sender's goal, and seeking the
principle themes of the message.

9.4.6.1 Achieving Effective Circular Communication

Achieving effective circular communication is through the perception
of the other person's emotions and to honestly reflect that. Taking the

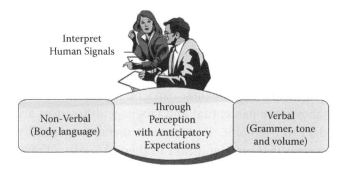

FIGURE 9.27
Interpreting human signals

responsibility of the success of the communication entails from the project manager to:

- Establish harmony: Demonstrate a body language compatible to the communication
- Engage: Invite to initiate the communication
- Encourage: Pursue the communication with minimal signals
- Be silent: Allow the other to reflect on what to communicate
- Question: Know when to ask open or closed questions
- Support: Demonstrate interest
- Feedback: Reformulate what is heard/seen

The project manager can maintain and sustain effective communication by proactively planning the methods, means, and frequency of exchanges with all project participants. This hinges on the predefined project communications plan, and goes beyond it by exploring further the contents and contexts of the messages. To achieve this, the project manager can determine how best to:

- Communicate frequently and target the audience
- Make the communications pertinent to the audience and the topic
- Establish specific communications for specific stakeholder groups at specific points in time
- Address the audience with the correct language and level: Avoiding buzzwords, three-letter acronyms where possible, esoteric terminology, seek credibility by steering clear of emotional and dramatic phrases

9.4.6.2 *Major Causes for Communication Breakdown*

There are several barriers that affect the flow of communication in a project. These barriers interrupt the flow of communication from the sender to the receiver, thus making communication ineffective. It is essential for project managers to overcome these barriers.

As seen above, the perceptual and language differences are the primary barriers to overcome. Project team members generally want to receive messages that are significant to them. Messages that are either impertinent or against their values may be ignored, rejected, and not accepted.

Projects by definition are under pressure to be completed within a specified time period within a constraining environment of scope expectations, funding, resource scarcity, risks, and changes. In the face of these adversities, and in a haste to meet deadlines, formal channels of communication are often shortened or bypassed, messages are partially given, emotions distort contents of decisions and instructions are hurriedly dispatched. All these factors contribute to ineffective communications.

Project communication will mostly flow along formal organizational lines within the project and the performing organization. The more there are levels of hierarchy and/or stringent communication rules, the greater is the probability that communication will be distorted or be diminished.

The checklist below will assist the project manager to preempt communication breakdowns:

- Bad communication skills
- Bad intentions
- Climate of fear
- Cultural differences
- Different goals
- Different level of knowledge
- Information "hiding"
- Lack of cultural awareness
- Lack of energy, sleep, food
- Lack of leadership
- Lack of respect
- Lack of structure
- Lack of trust

- Language differences
- Misunderstanding
- Technical problems
- Unaligned expectations
- Unclear agenda
- Unclear direction and goals
- Unclear priorities
- Unclear roles and responsibilities

9.4.6.3 Exercise: Communication Effectiveness

Reflect on your own communication effectiveness and identify the areas of improvement. Establish a personal action plan to put into practice on your current and future projects.

9.5 OVERVIEW OF CONFLICT MANAGEMENT IN PROJECTS

For any project to be effective and efficient in accomplishing its goals, the project manager and the team members, and other project players in the organization, need to have a shared vision of what they are striving to achieve.

The project manager needs to quickly recognize and resolve conflicts arising in the project, so as to avoid the problem or issue escalating to a level where cooperation between team members becomes difficult and/or impossible (see Figure 9.28).

Conflict can be defined in many ways:

- A struggle or clash between opposing forces; battle
- A state of opposition between ideas, interests, etc.; disagreement or controversy

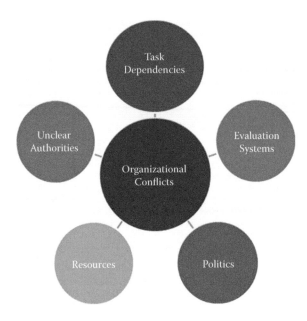

FIGURE 9.28
Types of organizational conflict

- A clash, as between two appointments made for the same time
- (Psychology) opposition between two simultaneous but incompatible wishes or drives, sometimes leading to a state of emotional tension and thought to be responsible for neuroses
- To come into opposition; clash
- To fight

There are differences between competition and conflict, where competition usually brings out the best in people, as they strive to be top in their field. Conflict is a situation of competition between two parties that are aware of the incompatibility of their respective positions and in which each party wishes to achieve a result that is irreconcilable with the desires or goals of the other.

There are opposing views on conflict, from the traditional view and belief that all conflict is harmful and must be avoided, to the human relations view and belief that conflict is a natural and inevitable outcome in any group. Additionally there is the view and belief that conflict is not only a positive force in a group but that it is absolutely necessary for a group to perform effectively.

9.5.1 Functional Conflict

Functional conflict is considered constructive as it can aid in developing individuals and improving the project performance by building on the individual assets of the team members. It may compel the project team to confront possible defects in a solution and choose a better one.

Functional conflict supports the goals of the group and improves its performance as it increases information sharing and ideas. It additionally encourages innovative thinking, promotes different points of view, and reduces stagnation. Project team members grow personally from the conflict and cohesiveness is strengthened among them.

9.5.2 Dysfunctional Conflict

Dysfunctional conflict is considered deconstructive when a problem remains unresolved, and the lack of decision is detrimental to the project.

Dysfunctional conflict hinders group performance, reduces trust, and creates tensions and stress. Due to poor decision making, there is excessive focus on the conflict and few new ideas are generated, which compounds the conflict.

9.5.3 Project Conflict Situations

Project conflict situations are principally divided into two categories: the functioning of the project or the functioning of the team participants. These categories are not clear-cut, as project conflicts often can create team conflicts and vice versa.

9.5.3.1 Functioning of the Project

- Scarcity of resources (finance, equipment, facilities, etc.)
- Disagreements about needs, goals, priorities, and interests
- Poor or inadequate organizational structure
- Lack of clarity in roles and responsibilities
- Conflicts over content and goals of the work

9.5.3.2 Functioning of the Team

- Different attitudes, values or perceptions
- Lack of teamwork

- Poor communication
- Conflict over how work gets done
  - At functional interfaces
  - Between managers and subordinates
  - Between departments
- Conflict based on interpersonal relationships
  - Between two or more people
  - Difference in ways on what should be done
  - Cultural
  - Threat to value system
- Unfair treatment

9.5.4 Conflict Resolution Techniques

The handling of conflict requires awareness of its various developmental stages. Project managers need to identify the conflict issue and how far it has developed, in order to solve it before it becomes a dysfunctional conflict.

The project manager can utilize different strategies to address a conflict, during which the power to influence exercised by each party will dictate the probable outcome of the resolution. The project manager's aim is to contain the issue to a functional conflict, where all parties seek an appropriate solution for the overall benefit of the project. This leads to a collaborating integrative negotiation. However, too often the ego of either or both parties "contaminates" any attempt to a resolution and the conflict becomes dysfunctional. Both parties must make an effort to separate the problem from the person.

The next sections describe the influence, negotiation, integrative and distributive negotiation strategies.

9.5.4.1 Understanding the Power Model

As a primer to conflict resolution techniques, it is wise to understand the power model as developed by French and Raven. Processes of power are pervasive, complex, and often disguised. They are however utilized in all conflict situations to influence the decision making (see Figure 9.29).

- The Bases of Social Power:
  - **Reward power:** based on the perceived ability to give positive consequences or remove negative ones

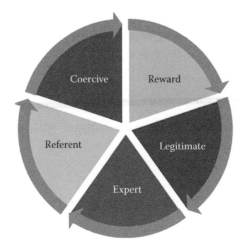

FIGURE 9.29
Power model

- **Coercive power:** the perceived ability to punish those who not conform with own ideas or demands
- **Legitimate power** (organizational authority): based on the perception that someone has the right to prescribe behavior due to election or appointment to a position of responsibility
- **Referent power:** through association with others who possess power
- **Expert power:** based on having distinctive knowledge, expertness, ability or skills
- **Information power (extension of expert power):** based on controlling the information needed by others in order to reach an important goal
- **Additional bases are:** Charismatic and Bureaucratic

For any type of power, the referent power will have the broadest range, while both expert and charismatic powers will personalize the attraction to the influencer. Each time reward or coercive power is used; a new state affects the relative positions of the two individuals. Reward power tends to result in increased attraction between the two individuals and lower resistance. Coercion results in a decreased attraction between the power individual and the follower and increased resistance. The more legitimate the coercion, the less it will produce resistance and decreased attraction.

For the other three main types of power, the relationship depends on the individual having the greater power—and this remains so after an influencing episode.

9.5.4.2 Assertive or Cooperative Techniques

In conflict management, the project manager will focus on two axes: Assertive or Cooperative.

- Assertive focus
  - Attempting to satisfy one's own concerns
- Cooperative focus
  - Attempting to satisfy the other party's concerns

The model below, sometimes called the win-win model, illustrates the different areas where the conflict can be resolved most appropriately, by using the proper amount of assertiveness and cooperation depending on the prevailing conditions (see Figure 9.30).

The interpretation of the different areas is:

- Collaborating (win-win)
  - This is a situation in which the parties to a conflict each desire to satisfy fully the concerns of all parties. The parties will try to

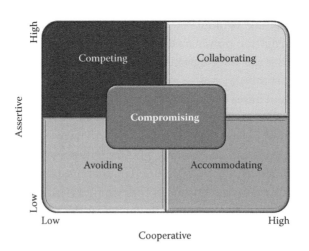

FIGURE 9.30
Key model for conflict management

handle the conflict without making concessions, by imagining a new way to resolve differences that are more beneficial to both.
- When best to use:
  - To find an integrative solution for an issue that cannot be compromised
  - To combine insight with different viewpoints
  - To gain commitment
  - To repair a relationship
  - Learning
- Compromising (partial win-win)
  - This is a situation in which each party to a conflict is willing to renounce something, and give way to their own goal. Each party seeks to accomplish a common goal and is willing to engage in finding a middle ground to reach a reasonable solution.
  - When best to use:
    - Goals are important but more assertive approaches may create a potential disruption
    - To reach provisional agreement to complex issues
    - To arrive at a decision under time pressure
    - Opponents with equal power are seeking mutually exclusive goals
    - To fall back when collaboration or competition is unsuccessful
- Competing (win-lose)
  - This is a situation when one party desires to satisfy its own interests, regardless of the impact on the other party to the conflict. Each party tries to maximize its own gain and has little interest in understanding the other's position.
  - When best to use:
    - Quick, decisive action is vital: in emergencies; on important issues
    - Unpopular actions need implementing (in cost cutting, enforcing unpopular rules, discipline)
    - On issues vital to the organization's welfare
    - "You know you're right"
- Accommodating (lose-win)
  - This is the situation when one party in a conflict places the other party's interests above their own. One party willingly gives in to the other party.

- When best to use:
    - Harmony and stability are especially important
    - Issues are more important to others and to maintain cooperation
    - To allow a better position to be heard
    - To build social credits for later issues
    - To minimize loss when "you're wrong," "outmatched," and "losing"
- Avoiding (lose-lose)
    - This is the situation when the two parties attempt to ignore the problem and do nothing to resolve the disagreement. Both parties have a desire to withdraw from or suppress the conflict.
    - When best to use:
        - Information is required before immediate decision
        - Issue is insignificant, or more important issues are to be addressed
        - Potential disruption outweighs the benefits of resolution
        - Others can resolve the conflict effectively
        - Issue seems indicative of other issues
- To regain composure

9.5.4.3 Strategies for Exercising Power

The French-Raven model provides the project manager with a palette of influence powers that can guide to an acceptable conflict resolution. It must be noted, however, that in certain cases the project manager must adopt a type of influence that may be uncomfortable, but is necessary and/or appropriate for the situation. As a basis, the manager must strive to provide impartial and objective information that causes others to feel the manager's course of action is correct. When the project manager's utilizes expert power, this lends credibility and ensures that everyone whose support is needed benefits personally from providing that support.

9.5.4.3.1 Influence Strategies
- Reason
    - This is the strategy most often used to persuade, relies on data and information and uses facts and logical arguments. It implicates planning, preparation, and know-how.

- Friendliness
  - This strategy rests on personality, relational capacities, and sensitivity in recognizing the mood and attitude of others. It consists in showing interest, goodwill, and respect to create a favorable impression.
- Consensus
  - This strategy mobilizes and rallies others to the reasoning. Influencing by the use of alliances and social networks with others.
- Interpersonal
  - A strategy that rests on the social norms of obligation and reciprocity, and places the emphasis on the exchange of benefits or favours. It is the art to influence others by negotiation.
- Directive
  - A strategy that involves giving orders and setting deadlines to accomplish the set desires, and demonstrating that the influencer is "in command."
- Hierarchical
  - This strategy utilizes the formal organizational lines to achieve a goal.

9.5.4.3.2 Negotiation Strategies

Negotiation is when parties to a conflict seek a solution acceptable to both, by considering various alternative ways to arrive at an agreement. Each party will impress its assertiveness or collaboration preferences. The strategies chosen can be either integrative negotiation or distributive negotiation. In some cases, both parties may arrive at a **B**est **A**lternative **T**o a **N**egotiated **A**greement (BATNA), which is the lowest acceptable value (outcome) to an individual for the negotiated agreement.

- Integrative negotiation—collaborating (win-win)
  - Negotiation that seeks one or more settlements—creates a win-win solution
  - Parties perceive that they might be able to achieve a creative solution to the conflict
  - Parties utilize collaboration or compromise
    - Focusing on the problem, not the people
    - Focusing on interests, not demands
    - Creating new options for joint gains
    - Focusing on what is fair

- Distributive negotiation—competing (win-lose)
  - Negotiation that takes a competitive adversarial stance—forces a win-lose situation
  - Parties see no need to interact in the future
- Parties do not care if their interpersonal relationship is damaged by their competitive negotiation

9.5.4.3.3 Approaches and Remedies for Difficult Situations

Difficult situations abound, and the project manager must be flexible to adapt a specific influence strategy to a specific person. The influence strategy is modified depending on the other party: team member, sponsor, stakeholder, peer or external provider. The project manager should recognize and utilize the influence techniques that have the most impact, but should not always use the same influence strategy with the same party. Ineffective influence strategies should not be pursued.

Certain influence strategies make project managers feel uncomfortable, such as: authoritarian, bargaining, coalition, etc., and they shy away from utilizing them. This is often the case when confronted with refusal and the project manager abandons easily to the conflict resolution. The project's objective is the ultimate goal of conflict resolution, and requires many times a temporary change of behavior for the situation on hand.

As the project manager develops exposure and experience in conflict management, there is a notable behavior change, which is a collection of the following characteristics:

- Good listener
- Sensitive to the needs of others
- Compassionate and understanding
- Emphasis on harmony
- Cooperative and not overly competitive
- Rapport building through conversations
- Advocating participation
- Communicating a desire to work together to explore a problem or seek a solution
- Seeking compromise rather than dominating
- Avoiding feelings or perceptions that imply the other person is wrong or needs to change
- Treating the other with respect and trust

9.5.4.4 Exercise: Conflict Management

Reflect on your project conflict resolution abilities and capabilities and determine areas to improve.

Index